Gödel

Gödel

A Life of Logic

John L. Casti
and
Werner DePauli

PERSEUS PUBLISHING

Cambridge, Massachusetts

Library of Congress Card Number: 00-105215
ISBN 0–7382-0274-6

Perseus Publishing is a member of the Perseus Books Group.

Find us on the World Wide Web at http://www.perseuspublishing.com

Perseus Publishing books are available at special discounts for bulk purchases in the U.S. by corporations, institutions, and other organizations. For more information, please contact the Special Markets Department at HarperCollins Publishers, 10 East 53rd Street, New York, NY 10022, or call 1–212–207–7528.

First printing, August 2000

1 2 3 4 5 6 7 8 9 10—03 02 01 00

Contents

Preface

As part of its millennium celebration, *Time* magazine published a list of the 100 greatest people of the 20th century. On this list was their choice of the greatest mathematician—Kurt Gödel. Most likely, if you were to randomly select 100 people and ask them, "Do you know who Kurt Gödel is?", it's almost certain you'd receive not a single positive response. That would definitely not be true if you asked about the greatest physicist of the 20th century (Einstein) or the greatest chemist (Linus Pauling?) or the greatest writer (??). This short volume is an attempt to bring Gödel's work and life to the attention of a broad audience, and thus to at least partially bring his magnificent achievements out into the mainstream of modern intellectual discourse.

This project had its beginning in 1986 with a television special on Gödel's life produced for the Austrian national television network by one of us (WDP) together with Peter Weibel. Following the screening of that program, a small book was published in German based on the script of that show. Initially, we thought to simply translate that book into English, slightly embellished with a bit more detail on Gödel's life and more references for the English reader. But it soon became evident that a more substantive volume was needed, and a totally new book emerged. That is the volume you hold in your hands.

We benefited greatly from the advice and assistance of a number of friends and colleagues in Vienna during the preparation of this manuscript. Specifically, we thank J. Golb, E. Köhler, C. Nielsen, and P. Weibel in this regard. In addition, the book's two editors, Jeff Robbins and Amanda Cook, have helped ensure that the book is readable, not an easy task when dealing with abstract work of the Gödelian kind.

Kurt Gödel may or may not have been the greatest mathematician of the 20th century; but he was certainly among them. It is our fondest hope that by the time the reader gets to the end of this book, he or she will know why.

JC
WDP

Chapter One

Since Aristotle

In 1965, the esteemed Austrian economist Oskar Morgenstern wrote to Austria's Foreign Minister (and later Chancellor) Bruno Kreisky that

> There is absolutely no doubt that Gödel is the world's greatest living logician; indeed, eminent thinkers such as Hermann Weyl and John von Neumann have declared that he is definitely the greatest logician since Leibniz, or even since Aristotle. It would seem that in the entire history of the University of Vienna the name of no figure teaching there has overshadowed Gödel's . . . . Einstein once told me that his own work no longer meant much to him, and that he simply came to the Institute to have the privilege of walking home with Gödel.

So who is this giant among giants that Morgenstern could claim was the greatest logician since Aristotle and that the great Einstein regarded as an intellectual peer? The answer will unfold in the pages of this book. But first, to get a glimpse of the work that would inspire such rhapsodies, let's take a quick look at what Kurt Gödel achieved in the worlds of both mathematics and philosophy.

Humans have always hungered for certain knowledge, the kind that transcends the millennia. We most assuredly can't find that kind of knowledge in the natural sciences, where even theories as fundamental as Newton's laws of mechanics can be overthrown by relativity theory, which itself may be cast in doubt by observations and experiments yet to come. Thus it is always to mathematics, especially the realm of pure numbers that we turn for the kind of certainty that we can really count on, if you'll pardon the poor pun. In this domain, the truth-generating mechanism we employ is the process of logical deduction bequeathed to us by Aristotle.

Basically, Aristotelian logic rests on two pillars: a set of *premises,* or *assumptions,* which are taken to be true without proof, and a collection of *rules of inference,* by which we transform one true statement into another. For example, consider the classical Socratic syllogism.

Premise A: All men are mortal.

Premise B: John is a man.

Conclusion: John is mortal.

Here premises A and B are assumed to be true statements about men, mortality, and a particular man, John. The leap from the two premises to the conclusion is taken by invoking one of the laws of deductive inference first spelled out by Aristotle: "If all X are Y, and Z is an X, then Z is a Y." As long as we can convince ourselves that the premises are true, then the conclusion that John is mortal is as solid and inescapable a fact as there ever will be. It is a certainty that follows from the semantic content of the premises and from the process of deduction laid down in our minds and formalized by Aristotle.

What Gödel discovered is that even though there exist true relationships among pure numbers, the methods of deductive logic are just too weak for us to be able to prove all such facts. In other words, truth is simply bigger than proof.

This fact doesn't seem too astonishing when put in the context of everyday life. We are all aware of things that we "know," but that we feel can never be logically deduced in a formal, Aristotelian fashion. In fact, the Oxford don and philosopher J. L. Austin, when first told of Gödel's result, remarked, "Who would have ever thought otherwise?" It seems likely that the average person on the street would say the same thing if someone announced that not everything that's true can be known by following a process of logical deduction. But not so for mathematicians! Mathematicians live in a world in which logical deduction is the very essence of their profession, and every accomplishment (theorem) that makes up the content of the practice of mathematics is the result of just such a chain of logical inference from propositions taken to be true without proof (such propositions are called axioms). Thus presenting incontrovertible proof that there are mathematical propositions that can be seen to be true but cannot be *proved* true, as Gödel did in 1931, hit the mathematical world like a wintry blast of arctic air. The pages that follow offer a rather leisurely, informal overview of Gödel's magnificent accomplishment and how he did it. But first, to put this work in context, it's helpful to glance briefly at the intellectual milieu in which Gödel lived and to look at the temper of the times that so crucially motivated his mind-bending results.

The Twilight of Monarchy

Brno, the city of the philosopher–physicist Ernst Mach, the novelist Gustav Meyrink, and the architect Adolf Loos, was the home of the Gödel family. At the turn of the century, this predominantly German-speaking city—now part of the Czech Republic—was the capital of Moravia and one of the Austro-Hungarian empire's most splendid urban centers. It was here that Gödel spent his childhood. Along with the centuries-old Jewish tradition of the Kabbala, with its mysticism and rigorous structure, the multicultural nature of the empire played an important role in the history of the Gödel family. In this culturally rich and diverse environment, Gödel was exposed to a wide variety of philosophical, artistic, and intellectual threads. One example of this influence is Gödel's interest in foreign languages. His papers contain notebooks on Italian, Dutch, and Greek as well as Latin; he spoke German, French, and English fluently, and his personal library contained many foreign-language dictionaries and grammars.

Gödel as a student (ca. 1930)

As a multiethnic state governed in a manner similar to the British Commonwealth, the monarchy suffered from an explosive nationality problem at that time—a problem rooted in events dating as far back as the fourteenth century. Not only was the mostly Hungarian region of Slovakia at odds linguistically with the Czech state and the German-speaking group of nations, but there were also increasingly common conflicts with Bohemia and Moravia, where, thanks to the Austrian ruling family, the German population had a significant influence. But the German nations were themselves politically fragmented—one portion favoring a rapprochement with Germany, another with Austria, which was still the dominant power in the southern German region.

The region's great spiritual tradition extended from the mysticism of Jakob Böhme (1575–1624) to the penetrating physical and philosophical analyses of Ernst Mach (1838–1916), founder of the analytic branch of the philosophy of science. Likewise, the tradition of Kabbalistic mysticism found its modern incarnation in both the surrealistic literature of Franz Kafka (1883–1932) and the portrayal of the "Golem"—the automaton brought to life in Jewish legend—by Gustav Meyrink (1868–1932).

Other noted intellectual figures from Brno and its vicinity included Gregor Mendel (1822–1884), the founder of genetics; and architect Josef Hoffmann (1870–1956), one of the founders of the *Wiener Werkstätte*, a movement that specialized in art deco design. By way of contrast, the architect Adolf Loos (1970–1933), also from Brno, advocated a form of austere constructivism, free of ornamentation, that anticipated Germany's Bauhaus movement.

Following the redrawing of the map of Europe after World War I a common thread running through the work of nearly all such intellectuals from the former Austro-Hungarian empire was the fusion of minute analysis with a mystical desire to transcend human boundaries. Because the cultural milieu is such a central factor in the unfolding of the creative imagination, the intellectual currents circulating in Brno at this time played a pivotal role in young Kurt Gödel's development.

This rich cultural climate was made possible by rapid industrial growth, particularly in the textile industry, which for the most part was in the hands of the predominantly Jewish, German-speaking population (Kafka's father worked in this industry). The prosperous industrialists lived in neighborhoods filled with spacious *Jugendstil* (literally: "youthful style," the German version of Art Nouveau) houses. But like all cities whose wealth comes quickly, Brno also had its dark side: impoverished areas that served both as a source of cheap labor and servants, mostly of Czech origin, and as a hotbed of crime.

After the collapse of the monarchy in the wake of World War I, the German-speaking minority in the new nation–state was faced with the difficult choice of either affiliating itself with Slavic culture, in order to avoid discrimination, or opting for association with Austria or Germany. Although the Gödel brothers had both learned a little Czech in school, their family origins were resolutely German. Thus they were predestined for a German-speaking cultural milieu, and they chose to attend university in Vienna instead of Prague.

Gödel's Intellectual Milieu

The period between World War I and World War II marked the peak of Vienna's most productive cultural period. In the autumn of 1924, Kurt Gödel settled in Vienna, the city of twelve-tone music, Jugendstil art, and the socio-linguistic critiques of Ludwig Wittgenstein and Karl Kraus, to continue his studies.

Philosophy, logic, and mathematics were the topics that absorbed the young Gödel. And as he moved deeper into these realms, he increasingly confronted the limits of language for expressing the truths of even such a pristine, simple realm as arithmetic, the relationships among numbers. Here Gödel was beginning to touch upon the work of the famed Vienna Circle.

One of the main goals of this group was to promote the ideas of Ernst Mach. Mach had argued that conceptually analyzing contradictions between theories leads to progress in our understanding of the natural world. His analytic method—presented in books such as *The History of Mechanics and Analysis of Sensations*—had fruitful consequences for both science and culture. In particular, by explaining and deconstructing physical concepts, Mach paved the way for Einstein's theory of relativity. His critique of Newtonian mechanics, with its notion of absolute time and space and the invariance principle of inertia, was of special historical importance. But Mach's theoretical approach also influenced artists, philosophers, and economists, thus shaping Austro-German discourse at the turn of the century. (Lenin feared Mach's influence on the cultural and political scene enough to publish in 1909 the polemic tract "Materialism and Empirico-Criticism," which ultimately denounced Mach's ideas.)

One of the outgrowths of The Ernst Mach Society, founded in Vienna in 1922, was the "Schlick Circle," named after its leading figure, Moritz Schlick, professor of mathematics at the University of Vienna and one of Gödel's philosophy teachers. With the appearance of its manifesto on "The Scientific World View" in 1928, the group became more publicly visible.

The Vienna Circle

If Plato's Academy in Athens served as the geographical focal point for Greek philosophy and its view of the world, then its twentieth-century counterpart can only be a small seminar room in the Mathematics Institute of the University of Vienna, where a group of physicists, mathematicians, and philosophers met every Thursday evening for several years in the 1920s and 1930s to debate the relationship between the theories of science and objective reality. This group, christened the Vienna Circle in 1931, eventually came to what amounts to the instrumentalist position that the only meaningful statements are those for which we can give a definite prescription (method or algorithm) for their verification. Thus, using a word like *yellow* would be equivalent to specifying a procedure for verifying that any particular object possessed the property of being yellow. In this way, the *meaning* or *reality* of "yellow" became equivalent to the *statement of the procedure for its verification.* This, in essence, forms the basis for the notorious Verification Principle, which lay at the heart of the school of *logical positivism,* the term later applied to the philosophy expounded by the Vienna Circle.

The triangle formed by language, world, and science was the focal point of the Vienna Circle's overall concerns. Thus it is not coincidence that a crucial aspect of Gödel's thinking was the limits of language. But to understand this blend of empiricism and logic, it is necessary to go back a few years and look at the work of Wittgenstein.

Wittgenstein and the Limits to Language

For ordinary men, the middle of a battlefield—with bullets flying and bombs bursting amid cries of human terror and agony—is hardly the kind of place in which to engage in contemplative philosophical speculation. But Ludwig Wittgenstein was no ordinary man, and during the course of his valiant service with the Austrian Army during World War I, he developed ideas about the relationship of thoughts expressed in language to the actual state of affairs in the world. These ideas were later enshrined in the pages of his classic work *Tractatus Logico-philosophicus*. The basic tenet of this seminal volume, which contains the only ideas of Wittgenstein's published during his lifetime, is that the structure of a sentence and the structure of the fact that the sentence asserts must have something in common. In this view, representation of the world in thought is made possible by logic, but the propositions of logic do not in and of themselves represent any actual state of the world. Thus logic is necessary but not sufficient to describe any kind of objective reality. However, for Wittgenstein, logic did reveal which states were theoretically possible, reflecting his underlying belief that reality was at least consistent. For example, if the statement "Water boils at 100°C at sea level" is true, then the statement "Water does not boil at 100°C at sea level" cannot also be true.

Wittgenstein's exploration of the interplay of language, logic, and observation of the world appealed to the members of the Vienna Circle, with their concerns about constructing a coherent philosophy of science from an amalgamation of logic and empirical epistemology. And indeed the *Tractatus* did serve as a point of departure for many of their deliberations, and several members of the circle were in regular contact with Wittgenstein in Vienna, although Wittgenstein himself seems never to have participated in the Thursday night discussions.

Wittgenstein's main conclusion—that language cannot capture all there is in the world—was given mathematical form by Gödel's work, as we'll see in detail in the next chapter. Essentially, what Gödel showed is that no kind of mathematics is ever going to be comprehensive enough to express fully the everyday notion of truth; in linguistic terms, no amount of syntax will ever entirely eliminate semantics. Even in the rarefied realm of pure numbers, formal logical operations will never tell us everything about the relationship among the integers. Again expressed linguistically, we must remember that arithmetic is *about* numbers. We'll see what this cryptic remark means in the next chapter.

Armed with this background material on the intellectual climate of Vienna in the 1920s, we are in a better position to understand the currents of thought in which Gödel's work was embedded. Let's now turn to what Gödel actually accomplished and see how and why his achievements immediately elevated him to the position of the greatest logician since Aristotle.

Chapter Two

Forever Incomplete

Demel's *Konditorei* is unquestionably Vienna's most famous pastry shop, celebrated by tourists and locals alike for its sinfully rich *Malakofftorte, Apfelstrudel,* and *Cremeschnitte,* all of which are served up smothered in that sine qua non of Viennese cuisine, vast expanses of whipped cream. What a boon to humanity it would be if someone were to invent a Chocolate Cake Machine (CCM) so that people everywhere could produce these goodies in the comfort of their own homes. What we have in mind is a device something like that shown in Figure 2.1.

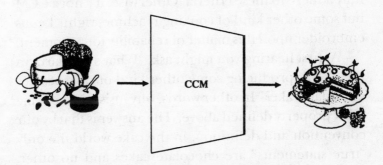

Figure 2.1. A Chocolate Cake Machine.

To operate the CCM we shovel into one end eggs, milk, flour, chocolate, and all the other raw materials that might be needed for a cake, along with a recipe for making, say, the famous *Sachertorte*. The CCM would then process the ingredients in accordance with the instructions given by the recipe, eventually serving up the *Sachertorte* at the machine's output slot. Of course, a sensible person would say that any decent pastry chef is a living embodiment of this CCM. But in the spirit of today's computer-oriented, high-tech culture, it's of more than passing interest to consider whether the entire process could be mechanized to the extent that it would be possible to build a CCM whose savory collations were indistinguishable from the chocolate cakes of the world's greatest pâtisseries. Let's think for just a minute about what such an ideal CCM should be like.

First of all, we want the CCM to be *reliable*. This means that when we feed in the ingredients and the recipe for something we are ready to label "chocolate cake," the CCM should produce chocolate cakes and nothing else. Of course, we need to have some criteria for what constitutes a chocolate cake. It is not the business of the CCM to establish these criteria, but once we have agreed upon a test for what does and what doesn't qualify as chocolate cake, the CCM should faithfully adhere to these criteria. Otherwise, it's not a CCM but some other kind of cooking machine, right? Let us embroider upon this matter of reliability for a moment.

At first hearing you might ask: "What's the harm in the CCM's producing some other kind of cake besides chocolate cake?" In other words, why insist on the reliability property defined above? The answer is that by our convention and definition, in the cake world the only "true statements" are chocolate cakes and no other. Therefore, because our interest is in having a machine that tells the truth and nothing but the truth, we are forced to demand that the CCM produce nothing but chocolate cakes. We can see the point more clearly if we generalize the situation and think in terms of a Universal Cooking Machine (UCM) instead of a CCM. The UCM is capable of producing any edible item, and in this more general world the true statements consist of anything and everything fit for human consumption. I think it's pretty obvious that we definitely don't want the UCM serving up inedible, or possibly even poisonous, items such as Strychnine Stew and Cobblestone Cobbler. Following the same line of reasoning, we don't want a Chocolate Cake Machine that serves up "false statements" like peanut butter cream cake and strawberry shortcake. But reliability is only half of what's needed for a good CCM.

If our contraption is to be a state-of-the-art, upmarket, yuppies-only CCM, then it should be possible to make any conceivable type of chocolate cake with it. Thus the CCM should be able to produce any object satisfying the chocolate cake test. In short, if it's chocolate cake, the CCM can make it. Let's call this the *totality* property.

These two properties—reliability and totality—constitute the basis for the kind of machine we all dream of having in our kitchen, ready for service at all times. Such a CCM would produce every possible type of chocolate cake, and only chocolate cake. But is such a CCM just an adolescent fantasy? Or can a sufficiently clever and motivated engineer with a sweet tooth duplicate Demel's chocolate cake chef in a machine, at least in principle?

Although on the surface it may seem far removed from any deep philosophical or epistemological considerations about life, the universe, or anything else, the question of the constructibility of a reliable and total CCM completely captures one of the most basic questions of philosophy and science: Is it possible to prove every truth? Or, to put it another way, is there any difference between a statement's being true and its being provable? Our goal here is to show that the answer to this question ultimately reduces to the seemingly much simpler question: "Can we build a CCM?" To get a better feel for what's involved in answering this foundational question, let's examine the issue in a bit more detail.

At those moments when we pass by Demel's window, our world consists solely of cakes—period. And if our particular interest in that world centers on a generous slice of *Sachertorte* smothered in whipped cream, then chocolate cake is the only brand of truth we recognize in this universe of cakes. Thus, in this cake context, the totality of all "statements" that can be made consists of a description of every conceivable type of cake. Some of these statements pass the test for being a chocolate cake and, hence, are "true." A description of any kind of cake that fails the test is relegated to the set of statements we label "false." Thus, for example, we call the statements *"Sachertorte," "Parisercremetorte,"* and *"Schwarzwälderkirschtorte"* true, while labeling the assertions *"Malakofftorte"* and *"Mohntorte"* false. It is evident that this cake world is like one of Plato's worlds of ideal forms: Its objects exist beyond the realm of space and time and are related to each other by sharing the abstract quality "cake." For us, the true statements in this world consist of those elements that pass the chocolate cake test. Some of these truths, such as *"Sachertorte,"* have actually been produced and may even be on display in Demel's window. But most have never appeared on the menu or in the window of any *Konditorei,* Viennese or otherwise. Consequently, the true statements of this cake universe consist of the totality of *all possible* chocolate cakes—real or only imagined.

Suppose now that you say that the statement *"Sachertorte"* is a true assertion in the cake world—that is, a *Sachertorte* satisfies our agreed-upon test for chocolate cakehood. A skeptic might reply, "I don't believe you. Prove it." How would you go about convincing a doubting Thomas of the correctness of your claim? What means would you employ to demonstrate that *"Sachertorte"* is a genuine truth of the cake world, meeting the stringent requirements of the chocolate cake test? The obvious answer: Just write down a recipe for *Sachertorte,* feed it into the CCM, and actually produce a real cake satisfying the criteria for chocolate cake. In the cake world, then, as in the rest of life, the proof is in the eating—literally! A statement (a cake) is provable (is a genuine chocolate cake) if and only if there is a recipe that can be followed by the CCM for actually making that cake. But note carefully that for a cake to be provable, it's not necessary actually to *implement* the recipe with the CCM. It suffices to provide the recipe and show that if you *did* feed it into the CCM, the result would indeed be something satisfying the chocolate cake test. In other words, provability means that there is a rule that could be followed to produce the cake. Thus, for the universe of cakes, we have

Truths = all conceivable cakes satisfying the chocolate cake test

Proofs = all recipes for actually making chocolate cakes with the CCM

Now comes the Big Question: Is there a recipe for every conceivable chocolate cake? Or, equivalently, is every true statement provable? What we're asking here is whether there are honest-to-god chocolate cakes in the Platonic universe of cakes for which no recipe can ever be given. Or can every object that satisfies the chocolate cake test actually be produced by following a set of instructions? Looked at from the perspective of the CCM, the question reduces to whether there is any theoretical barrier to the construction of a reliable and total CCM?

Most pastry chefs, amateur or otherwise, would probably answer that if you can imagine it, you can not only make it, but also write down the recipe so that anyone else can make it, too. Interestingly enough, until 1931 not only pastry chefs but just about every mathematician would have agreed with this claim. But believing and knowing are radically different matters, and in that fateful year, Kurt Gödel showed conclusively that what's true and what's provable are just not the same thing at all—and not only in the restricted universe of cakes. Gödel's remarkable result, which many regard as the most profound and far-ranging philosophical result of this century, applies to the vastly broader universe of general, everyday events.

Stripped to its bare essentials, what Gödel's Theorem did was shatter forever the belief that there is no difference between truth and proof. The theorem's punch line is that there is an eternally unbridgeable gap between what's true (and can even be seen to be true) within a given logical framework or system and what we can actually prove by logical means using that same system. Hence, despite the best efforts of an army of chefs, the "cake bibles" are forever doomed to incompleteness; there will always exist chocolate cakes that can be seen to be bona fide chocolate cakes, yet whose recipes can never be written down.

Now how did Gödel actually *prove* such a result about the limits to proof? That's a story that will take a few pages to tell.

The Limits to Proof

Roma, Venezia, Milano, Firenze, Napoli—*de rigueur* ports of call on the "See Italy in Five Days" package tours. And that's as it should be for Cousin Katy from Kankakee on her once-in-a-lifetime pilgrimage to the land of Benetton, the Mafia, and Leonardo. But those jaded travelers looking for just a bit more than the obligatory churches, statues, and museums will have none of the blandishments of the tour operators, heading instead for the exit as the train pulls into the Bologna station midway between Venice and Florence. Yes, Bologna. For besides being regarded by many gourmets as the eating capital of Italy—a sort of Italian counterpart to the French culinary mecca of Lyons—Bologna is also the focal point of the Italian exotic sports car industry. The Lamborghini, Ferrari, Maserati, and de Tomaso factories are all located within a few miles of beautiful (really!) downtown Bologna. As though this were not enough, Bologna also claims the distinction of having the world's oldest university. And it was at this venerable site, during the 1928 International Congress of Mathematicians (ICM), that the famed German mathematician David Hilbert threw down a challenge that would forever change the way we think about the relationship between what is logically provable and what is actually true.

At stake in Hilbert's 1928 address was the foundational issue of whether it's possible to prove every true mathematical statement. What Hilbert was looking for was a kind of Truth Machine. Just feed the statement in at one end, turn the crank, and sit back as out the other end pops the answer: TRUE or FALSE. Ideally, in this setup the original statement would be either a true mathematical fact and, hence, logically deducible from the given assumptions and thus a theorem, or it would be false and, consequently, not a theorem—that is, its negation would be a theorem. In short, Hilbert's Truth Machine would give a complete account of every mathematical assertion. In his Bologna talk, Hilbert laid down the requirements for such a Truth Machine, or what is more pedantically termed an *axiomatic,* or *formal, logical system,* along with the conviction that his "Program" would ultimately yield a complete axiomatization of all of mathematics.

With this challenge to the mathematical world, Hilbert was reemphasizing a different aspect of another problem he posed at an earlier ICM gathering in Paris in 1900. Because unsolved problems are the lifeblood of any field of intellectual activity, to mark the turn of the century, Hilbert listed 23 problems whose resolution he felt was of crucial importance for the development of mathematics. The second problem on this list involved proving that mathematical reasoning is reliable. In other words, by following the rules of mathematical reasoning, it should not be possible to arrive at mutually contradictory statements; a proposition and its negation should not both be theorems. Of course, this self-consistency requirement is a necessary condition for any axiomatic system of the sort Hilbert had in mind; if the system is inconsistent, it's possible to prove any assertion TRUE or FALSE as we wish—hardly a secure basis for reliable knowledge.

As an amusing illustration of the crucial importance of self-consistency, Bertrand Russell once gave the following "proof" that if $2 + 2 = 5$, then he was the Pope. Here's Russell's argument: If we admit that

$2 + 2 = 5$, then we can subtract 2 from each side of the equation, giving us $2 = 3$. Transposing, we have $3 = 2$, and subtracting 1 from each side of this equation gives us $2 = 1$. Thus, because the Pope and Russell are two people and $2 = 1$, the Pope and Russell are one. Hence Russell is the Pope! This is about as good an argument as any for why an inconsistent logical system is basically useless in terms of getting at the truth.

But why was Hilbert even concerned in the slightest about such matters? After all, at least since the time of Euclid, mathematicians had been using successfully the very methods that worried Hilbert. Why get worried now? Was $2 + 2$ all of a sudden going to become 4.007? Or was the sum of the angles of a triangle going to turn out to differ from 180 degrees? Actually, it was exactly this question about triangles that served as one of the sparks that ultimately touched off Hilbert's concern. In the early part of the nineteenth century, the geometers János Bolyai and Nikolai Lobachevski had shown independently, and quite contrary to popular belief and everyday intuition, that there were other perfectly consistent ways of mathematically talking about things like points and lines besides the way of Euclid. And in these "noneuclidean geometries," the sum of the angles of what passes for a "triangle" could be less than 180 degrees (hyperbolic geometry) or greater than 180 degrees (elliptic geometry).

Thus, despite its unquestioned utility in the physical world, Euclid's geometry turns out to be no more or less "true" than its competitors, at least in the universe of mathematical objects. And, in fact, even in the physical world these noneuclidean geometries come into their own when we start considering objects on a cosmological scale. For example, on the basis of current observations of the distribution of matter in the universe, it's beginning to look more and more likely that the large-scale structure of the universe obeys the geometry of Bolyai and Lobachevski, in which, given a fixed line and a point not on that line, we can draw an infinite number of lines through the point, all of which are parallel to the given line. This is in stark contrast to the world of Euclid, in which only a single such parallel line may be drawn.

Alternative geometries call into question the relationship between mathematical objects and the external world, because by definition the universe is the real world, whereas points, parallel lines, and triangles seem to be far less tangible, existing as much in the mind as in the universe of material objects and everyday events. But far more troubling to Hilbert than noneuclidean geometries were the logical paradoxes discovered by Bertrand Russell and his followers shortly after the turn of the century. These logical puzzlers are exemplified by the famous Barber Paradox: "The village barber shaves all those in the village who do not shave themselves. Who shaves the barber?" Tracing through the logical possibilities, we find that if the barber shaves himself, then he doesn't shave himself—and vice versa.

The standard methods of logical inference are too feeble to settle even such a seemingly simple question as the Barber Paradox. Nevertheless, these are precisely the tools upon which the methods used in constructing mathematical proofs ultimately rest, and this suggests why Hilbert and others started getting concerned about the logical coherency of the mathematical enterprise. In Hilbert's own words, "Every definite mathematical problem must necessarily be susceptible of an exact settlement, either in the form of an actual answer to the question asked, or by a proof of the impossibility of its solution." But within the framework of classical logic, the Barber Paradox is just plain undecidable. Therefore, Hilbert was challenging his colleagues to formalize every mathematical truth in a way that would forever exclude the possibility of paradoxical statements appearing in mathematics as Russell had showed that they could appear in ordinary language and logic.

The outcome, however, was something else entirely. Less than three years after Hilbert's Bologna address, the young Austrian logician Kurt Gödel astonished the mathematical world by publishing a revolutionary paper transforming Hilbert's fondest dream into his worst nightmare. Before turning to an account of these matters, let's first see how a statement of mathematics, of arithmetic even, can be true—but unprovable in a mathematical sense.

According to mathematics folklore, one day during the very brief grade school career of Karl Friedrich Gauss (1777–1855) (arguably the greatest mathematician of all time), Gauss's teacher grew annoyed with his students' unruly behavior and decided to silence them for a while by assigning a long calculation to perform. Specifically, the teacher told the class to add up all the numbers between 1 and 100. By the custom of the time, the first student to finish was supposed to write the answer on a slate and then put the slate face down on the teacher's desk. Entertaining the happy vision of long columns of numbers and frequent childish calculational errors, the teacher no doubt felt that this chore would occupy the class long enough for him to regain his sanity and peace of mind. Unfortunately, he hadn't counted on having a mathematical prodigy in the room, and within a few moments after the problem had been given, Gauss's slate slammed down on his desk. How did Gauss do it?

Being considerably more clever than the teacher, Gauss saw immediately that the way to solve this problem was to separate the numbers from 1 to 100 into two groups and then write these two groups one below the other in the following way:

$$1 \quad 2 \quad 3 \quad 4 \quad 5 \quad \cdots \quad 50$$
$$100 \quad 99 \quad 98 \quad 97 \quad 96 \quad \cdots \quad 51$$

What Gauss noticed was that when he added the corresponding numbers from each group, the sum was always the same, 101. And because the one hundred numbers were divided into two groups of equal size, there had to be 50 such pairs. Consequently, the sum of the numbers from 1 to 100 must be equal to 50×101, or 5050. And it doesn't take much exercise of the imagination to see that Gauss's trick will work for any number the teacher might have chosen. If the teacher had given the number n, you just separate the integers from 1 to n into two groups of equal size (0 has to be thrown in to balance the groups if n happens to be odd) and then write the two groups in ascending and descending order as above. The desired sum, $1 + 2 + 3 + \cdots + n$, will then equal $(n/2) \times (n + 1)$.

Gauss's scheme constitutes a proof of this formula for an arbitrary, *but fixed*, whole number n. It is not a proof that the formula holds for *every* positive integer n; it's just a proof for any fixed number the teacher may happen to call out. The usual proof of the general formula makes use of the principle of mathematical induction. We first verify that the formula holds for the case $n = 1$. Next we assume that it holds for an arbitrary, but fixed, positive integer n. We then use this assumption to *deduce logically* that it holds for $n + 1$. Thus we show that if it holds for $n = 1$, then it holds for $n = 2$. And if it holds for $n = 2$, then it holds for $n = 3$, and so on. In fact, *all* proofs of this basic formula of arithmetic make use of an inductive argument of this sort in one way or another. This technique of mathematical induction, though it is not a tool of formal logical inference, is used extensively in mathematical arguments to enable us to infer a result for an infinite number of cases (all positive integers) from a finite set of conditions (the two cases $n = 1$ and $n =$ arbitrary, but fixed).

There are some philosophers of mathematics who argue that such nonconstructive and/or infinitary principles of inference as mathematical induction should not be admitted into mathematics as a tool of proof. If, accordingly, we were to strip out the tool of induction from the logical proof mechanism of mathematics, the formula for the sum of the first n integers would no longer be provable for general n. Nevertheless, we would still be able to see "from the outside," so to speak, that the formula is true. What Gödel showed is that even when we can avail ourselves of all the tools of logical inference and mathematical proof, including mathematical induction, certain true mathematical statements remain unprovable. In short, there is an eternally unbridgeable gap between what can be proved and what's true.

As a consequence of Gödel's work, issues of prediction and explanation in mathematics center on the following foundational questions:

Question I: Proof vs. Truth
What are the limits to mathematical proof?

Question II: Mathematical Reality
What does a mathematical "proof" prove?

Question I addresses the degree to which we can hope

to narrow the gap between the universe of true mathematical statements and what's provable. But all mathematical truths, provable or otherwise, are statements about the existence of certain kinds of objects. What kind of an existence does something like a hyperbolic triangle or a random number actually have? Thus Question II asks us to give an account of what we are really talking about when we claim to have proved the "existence" of an object such as an elliptic triangle or the formula for the sum of the first n integers. In cakeworld terminology, we can think of these big questions as asking to what degree we can write down a recipe for every possible chocolate cake (Question I) and what kind of an existence a chocolate cake has if we give its recipe but have never actually baked it (Question II).

If our own existence were confined solely to the universe of cakes, these questions might seem to be at best amusing philosophical digressions. But unless you live in Vienna, your world is probably a bit broader in scope than the offerings of the corner coffee house and *Konditorei*. In particular, questions of prediction and explanation in the worlds of both science and mathematics ultimately come down to the kinds of answers we're able to provide for Questions I and II. Because the idea of an axiomatic framework for all of mathematics was the starting point for Gödel's assault on proof, let's begin our story with a bit of background on Hilbert's Program for axiomatizing mathematical truth.

Speaking Formally

In his famous epigram on the nature of mathematics, Bertrand Russell claimed that "pure mathematics is the subject in which we do not know what we are talking about, or whether what we are saying is true." This pithy remark summarizes the content of both Questions I and II, as well as cutting directly to the heart of Hilbert's Program: the development of a purely syntactic framework for all of mathematics. There's more than a touch of irony in Russell's remark asserting the content-free nature of mathematics, because a prime force motivating Hilbert's Program was his feeling that the paradoxical element in things like Russell's own Barber Paradox was due to the *semantic* content in the statement of the paradox. Hilbert believed that the way to eliminate the possibility of such paradoxes arising in mathematics was to create an essentially "meaningless" framework within which to speak about the truth or falsity of mathematical statements. Such a framework is now termed a *formal system,* and it constitutes the historical jumping-off point for investigations of the gap between what can be proved and what is actually true in the universe of mathematics.

The "meaningless statements" of a formal system are composed of finite sequences of abstract *symbols.* The symbols are often termed the *alphabet* of the system, whereas the "words" of the system are usually called *symbol strings.* The symbols might be objects like ♠ and ♡, or they might even be signs like 0 and 1. But in the latter case, it's absolutely essential to recognize that we're not talking yet about the *numbers* 0 and 1 but only about the *numerals* 0 and 1. It's only when these symbols are given meaning as numbers that they acquire the properties we usually associate with the numbers 0 and 1. We'll come back to this point with a vengeance shortly. In a formal system, a finite number of these symbol strings are taken as the *axioms* of the system. To round things out, the system also has a finite number of *transformation rules.* These rules specify how a given string of symbols can be converted into another such string.

The general idea of proof within a formal system is to start from one of the axioms and apply a finite sequence of transformations, thereby converting the axiom into a succession of new strings, where each string either is itself an axiom or is derived from its predecessors by application of the transformation rules. The last string in such a sequence is called a *theorem* of the system. The totality of all theorems constitutes what can be proved within the system. But note carefully that these so-called statements don't actually say anything; they are just strings of abstract symbols. We'll get to how the theorems acquire meaning in a moment. But first let's see how this setup works with a simple example.

Suppose the symbols of our system are the three objects ♠ (spade), ♡ (heart), and ♣ (club). Let the two-element string ♡♣ be the sole axiom of the system. Letting x denote an arbitrary finite string of spades, hearts, and clubs, we take the transformation rules of our system to be

Rule I:	x♣	$\longrightarrow$	x♣♠
Rule II:	♡x	$\longrightarrow$	♡xx
Rule III:	♣♣♣	$\longrightarrow$	♠
Rule IV:	♠♠	$\longrightarrow$	—

In these rules, $\longrightarrow$ means "is replaced by." For instance, Rule I says that we can form a new string by appending a spade to any string that ends in a club. The interpretation of Rule IV is that any time two spades appear together in a string, they can be dropped to form a new string. Now let's see how these rules can be used to prove a theorem.

Starting with the single axiom ♡♣, we can deduce that the string ♡♠♣ is a theorem by applying the transformation rules in the following order:

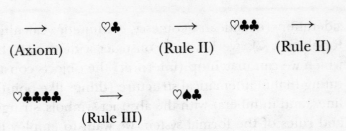

Such a sequence of steps, starting from an axiom and ending at a statement like ♡♠♣, is termed a *proof sequence* for the theorem represented by the last string in the sequence. Observe that, when applying Rule III at the final step, we could have replaced the last three ♣s from the preceding string rather than the first three, thereby ending up with the theorem ♡♣♠ instead of ♡♠♣. The perceptive reader will also have noted that all the intermediate strings obtained in moving from the axiom to the theorem begin with ♡. It's fairly evident from the axiom and the action of the transformation rules for this system that every string will have this property. This is a *metamathematical* property of the system, because it's a statement *about* the system rather than one made *in* the system itself. The distinction between what the system can say from the inside (its strings) and what we can say about the system from the outside (properties of the strings) is of the utmost importance for Gödel's results.

About now the right question to be asking yourself is "What does all this meaningless symbol manipulation have to do with everyday reality?" Let's quickly turn our attention from matters of form to those of content.

The answer to how we get from form to content can be given in one word: *interpretation*. In particular, for reasons that will become apparent in a moment, let's focus our interest right now on the slice of everyday reality consisting of mathematical facts. Then, depending on the kind of mathematical structure under consideration (euclidean geometry, arithmetic, calculus, topology, ...), we have to construct a dictionary by which we can match up (interpret) the objects constituting that mathematical structure (things like points, lines, and numbers) with the abstract symbols, strings, and rules of the formal system we want to employ to represent that structure. By this dictionary construction step, we attach meaning, or semantic content, to the abstract, purely syntactic strings formed from the symbols of the formal system. Thereafter, all the theorems of the formal system can be interpreted as true statements about the associated mathematical objects. The accompanying diagram illustrates this crucial distinction between the purely syntactic world of formal systems and the meaningful world of mathematics.

Before proceeding further, let's pause here for a moment to explain a point about this interpretation step that could puzzle the attentive reader.

Originally, Hilbert suggested the idea of a formal system for getting at mathematical truth as a way of eliminating the possibility that logical paradoxes of the barbering sort could stick their ugly, unshaven faces into the realm of mathematics. The main selling point for formalization was the claim that these kinds of paradoxes stemmed from the semantic content of their expression in natural language. Hence, if the symbols and strings of the formal system are completely meaningless, then the statements (symbol strings) should be paradox-free. In particular, there should be no undecidable propositions. But if that argument is the main selling point for formal systems, then why are we all of a sudden trotting out this interpretation step and thereby injecting meaning back into the picture? Doesn't this dictionary construction step undermine Hilbert's entire argument for formalization?

Formal World
(Syntactics)
symbols and strings
axioms
rules of inference
⇓
theorems

⇑

Dictionary

⇓

Mathematical World
(Semantics)
arithmetic (e.g., number theory)
geometry (e.g., topology)
analysis (e.g., calculus)
⇓
mathematical truths

The key to resolving this apparent dilemma lies in putting the horse before the cart. Hilbert's program involved *starting* with the formal system. The second step was then to bring out the mathematical structure of concern and show how to match its objects to the strings of the formal system—that is, how to interpret the meaningful mathematical objects in terms of the meaningless formal ones. Thus we don't begin with the semantic-laden mathematical structure but, rather, start with the purely syntactic world of the formal system. Hilbert's Program really amounted to trying to find a formal system that was above all free from internal contradictions and whose theorems were in perfect correspondence with all the true facts of arithmetic. In essence, Hilbert didn't believe that any Russell-type paradoxes lurked in the world of mathematical truths, even though they might exist in the far fuzzier realm of natural language. And the way he thought we could prevent them from crossing the border separating ordinary language from mathematics was to formalize the entire universe of mathematical truth. What Gödel showed was that Hilbert was dead wrong. There is simply no way to erect a barrier between mathematics and the demons of undecidability—even in the pristine, crystal-clear world of pure numbers. Now let's get back to our story.

Once the dictionary has been written linking a mathematical structure with a formal system and the associated interpretation has been established, we can hope along with Hilbert that there will be a perfect, one-to-one correspondence between the true facts of the mathematical structure and the theorems of the formal system. Loosely speaking, Hilbert's dream was to find a formal system in which every mathematical truth translates into a theorem, and conversely. Such a system is termed *complete*. Moreover, if the mathematical structure is to avoid contradiction, a mathematical truth and its negation should never both translate into theorems—that is, be provable in the formal system. Such a system in which no contradictory statements can be proved is termed *consistent*. With these preliminaries in mind, we can finally describe Gödel's wreckage of Hilbert's Program.

By the time of Hilbert's 1928 Bologna lecture, it was already known that the problem of the consistency of mathematics as a whole was reducible to the determination of the consistency of arithmetic—that is, the consistency of the properties and relations among the natural numbers (the positive integers 1, 2, 3, . . . , or what some people call the whole numbers). Therefore, the problem became to give a "theory of arithmetic," a formal system that was (1) finitely describable, (2) complete, (3) consistent, and (4) sufficiently powerful to represent all the statements that can be made about the natural numbers. What Hilbert meant by *finitely describable* was not only that the number and length of the axioms and rules of the system should be constructible in a finite number of steps, but also that every provable statement in the system (every theorem) should be provable in a finite number of steps. This condition seems reasonable enough, because you don't really have a theory at all unless you can tell other people about it. And you certainly can't tell them about it if there are an infinite number of axioms, rules, and/or steps in a proof sequence.

A central question that arises in connection with any such formalization of arithmetic is whether there is a finite procedure by which we can decide the truth or falsity of every arithmetical statement. For example, if we make the statement "The sum of two odd numbers is always an even number," we want a finite procedure, essentially a computer program, that halts after a finite number of steps, telling us whether that statement is true or false—that is, provable or not in some formal system powerful enough to encompass ordinary arithmetic. For example, in the ♡-♠-♣ formal system we considered earlier, such a decision procedure is given by the following far from obvious criterion: "A string is a theorem if and only if it begins with a ♡ and the number of ♣s in the string is not divisible by 3." The question of the existence of a mechanical procedure or rule like this to decide every statement about arithmetic is Hilbert's famous Decision Problem, which we already know is unsolvable.

Hilbert was convinced that a formalization of arithmetic satisfying the foregoing desiderata was possible, and his Bologna manifesto challenged the international mathematical community to find or create it. It's somehow comforting to know how dramatically and definitively wrong even a man as great as Hilbert can be!

In 1931, less than three years after Hilbert's Bolognese call to arms, Kurt Gödel published the following metamathematical fact, perhaps the most famous mathematical (and philosophical) result of this century:

Gödel's Theorem—Informal Version
Arithmetic is not completely formalizable.

Remember that for a given mathematical structure like arithmetic, there are an infinite number of ways in which we can choose a finitary set of axioms and rules of a formal system in an attempt to mirror syntactically the mathematical truths of the structure. What Gödel's result says is that *none* of these choices will work; there does not and cannot exist a formal system satisfying all the requirements of Hilbert's Program. In short, there are no rules for generating *all* the truths about the natural numbers.

Gödel's result is shown graphically in Figure 2.2 for a given formal system M representing arithmetic. The entire square represents all possible statements that can be made about the natural numbers. Initially, the square is entirely gray. When we prove a statement true by applying the rules of the formal system M, we color that statement white; when we prove a statement false, we color it black. Gödel's Theorem says that there will

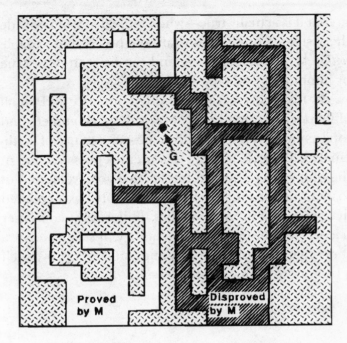

Figure 2.2. Gödel's theorem in logic space.

always exist statements (or *Gödel sentences*) such as G that are eternally doomed to a life in the shadow world of gray; it's impossible to eliminate the gray and color the entire square in black and white. And this result holds for *every* possible formal system M, provided only that the system is consistent; for every consistent formal system M, there is at least one statement G that cannot be proved or disproved in M. As it is in the rest of life, so it is in arithmetic too: There's no washing away the gray! We call a statement like G *undecidable* in M, because it can be neither proved nor disproved within the framework of that formal system. And if we add that undecidable statement G as an axiom, thereby creating a new formal system, the new system will have its own Gödel sentence. What's perhaps equally surprising is that for every such formal system M, the statement G can be constructed

so that it is actually true—when looked at from outside
the system. Consequently, although G is undecidable
within M, it is actually true and can be seen to be true
by jumping outside the system.

By his theorem, Gödel snuffed out once and for all
Hilbert's flickering hope of providing a complete and
total axiomatization of arithmetic—and hence of mathematics. Because Gödel's Theorem represents one of
the pinnacles of human intellectual achievement, not
to mention forming the basis for a whole host of related
developments in mathematics, philosophy, computer
science, linguistics, and psychology, it's well worth a
short chapter looking at how one could ever prove such
a profound, mind-boggling result.

Chapter Three

The Undecidable

Unfortunately, the full mathematical details of Gödel's Theorem are much too complicated for the type of presentation we're interested in here. But the basic ideas, though a bit tricky and devious, are still fairly easy to grasp even for those with no mathematical training. In fact, understanding the logic underlying Gödel's magnificent achievement has been described by some as being akin to a religious or mystical experience.

In arriving at his proof of the incompleteness of arithmetic, Gödel's first crucial observation was to recognize the importance of Hilbert's insight that every formalization of a branch of mathematics is itself a mathematical object in its own right. This means that if we create a formal system intended to capture the truths of arithmetic, that formal system can be studied not just as a set of mindless rules for manipulating symbols, but also as an object possessing mathematical—that is, semantic—as well as syntactic properties. In particular, because Gödel was interested in the relationships between numbers, he showed how it would be possible to represent, within arithmetic itself, any formal system purporting to encompass arithmetic. In short,

Gödel saw a way to mirror all statements about relationships between the natural numbers by using these very same numbers themselves.

This mirroring idea is probably more familiar in the context of ordinary language, where we use words *in* the English language to speak *about* language. For example, we use *words* to describe *properties of words,* such as whether they are nouns or verbs, and we discuss the structure of, say, a treatise on English grammar, which consists of words, by employing other words of the English language. Thus, in both cases we are making use of language in two different ways: (1) as a collection of *uninterpreted* strings of alphabetic symbols that are manipulated according to the rules of English grammar and syntax, and (2) as a set of *interpreted* strings that have a meaning within the context under discussion. The key notion is that the very same objects can be considered in two quite distinct ways, opening up the possibility for that object actually to speak about itself. In passing, we should note that the very same dual-level idea applies to the symbols and their interpretations in the genetic material (the DNA) of every living cell. What Gödel saw was how to perform this same "trick with mirrors" using the natural numbers.

To understand Gödel's "mirroring" operation with numbers a little more clearly, consider the familiar situation at a bakery counter where each customer is given a number upon entering in order to indicate the sequence in which the customers will be served. Suppose Clint and Brigitte both want to have a slice of *Sachertorte* and haven't yet obtained a CCM. Thus they go down to the local *Konditorei*, where, upon entering, Clint receives the service number 4; Brigitte comes in a bit later and gets number 7. By this service assignment scheme, the real-world fact that Clint will be served before Brigitte is "mirrored" in the purely arithmetical truth that 4 is less than 7. In this way a truth of the real world has been faithfully translated, or mirrored, by a truth of number theory. Gödel used a tricky variant of this kind of numbering scheme to code all possible statements about arithmetic using the language of arithmetic itself, thereby employing arithmetic both as an interpreted mathematical object and as an uninterpreted formal system with which to talk about itself. It's revealing to see how this *Gödel numbering* scheme actually works.

In their monumental three-volume treatise *Principia Mathematica,* Bertrand Russell and Alfred North Whitehead used the symbolism of logic to create a calculus capable of expressing the statements of arithmetic, geometry, and analysis—essentially all of classical mathematics. If you're ever tempted to dip into this work and see what "2 + 2 = 4" looks like in logical language, let me warn you that I once succumbed to the same temptation. I was ultimately buried in an impenetrable morass of abstract symbols and formulas that didn't get around to the proof of "1 + 1 = 2" until Volume Two! It's no wonder that philosopher and educator John Kemeny could describe the Russell–Whitehead work as "a masterpiece discussed by practically every philosopher and read by practically none." But for Gödel, the *Principia* was a convenient starting point for his tail-swallowing idea of turning arithmetic back upon itself by coding all the symbols and statements of the Russell–Whitehead language in arithmetic. In this way Gödel was able to describe the results on each page of *Principia Mathematica* as a sequence of transformations of numbers.

To see how Gödel's method works, let's consider a somewhat streamlined version of the Russell–Whitehead language of symbolic logic that we owe to Ernest Nagel and James R. Newman. In this language there are elementary signs and variables. To follow Gödel's scheme, suppose there are the ten logical signs shown in Table 3.1, each with its Gödel code number, an integer between 1 and 10.

Table 3.1. Gödel numbering of elementary logical signs.

Sign	Gödel Number	Meaning
~	1	not
∨	2	or
⊃	3	if … then
∃	4	there exists
=	5	equals
0	6	zero
s	7	the immediate successor of
(	8	punctuation
)	9	punctuation
′	10	punctuation

In addition to the elementary signs, the language of the *Principia* contains logical variables that are linked through the signs. These variables come in three different flavors, representing a kind of hierarchical ordering that depends on the exact role the variable plays in the overall logical expression. Some variables are *numerical,* which means that they can take on numerical values. For other variables (*sentential* variables), we can substitute entire logical expressions or formulas. And finally, we have what are called *predicate* variables, which express properties of numbers or numerical expres-

sions; examples include *prime, odd,* or *less than.* All the
logical expressions and provability relations in *Principia
Mathematica* can be written by using combinations of
these three types of variables and connecting them via
the logical signs. For our streamlined version of *Principia,* there are only ten logical signs, although in the
real case there are quite a few more. In this toy version of *Principia Mathematica,* Gödel's numbering system would code numerical variables by prime numbers
greater than 10, sentential variables by the squares of
prime numbers greater than 10, and predicate variables
by the cubes of prime numbers greater than 10.

To see how this numbering process works in prac-
tice, consider the logical formula $(\exists x)(x = sy)$. Translated into plain English, this logical formula reads,
"There exists a number x that is the immediate suc-
cessor of the number y." Both x and y are numerical variables, so the Gödel coding rules dictate that we
make the assignment $x \rightarrow 11$, $y \rightarrow 13$, because 11 and
13 are the first two prime numbers larger than 10. The
other symbols in the formula can be coded by substituting numbers using the correspondences in Table 3.1.
Carrying out this coding yields the sequence of numbers (8, 4, 11, 9, 8, 11, 5, 7, 13, 9), formed by reading
the logical expression symbol by symbol and substituting the appropriate number according to the coding
rule. This sequence of ten numbers pins down the logical formula uniquely. But because number theory—
that is, arithmetic—is about numbers, we'd like to be
able to represent the formula in an unambiguous way
by a single number. Gödel's procedure for doing this
is to take the first ten prime numbers (there are ten
symbols in the formula) and multiply them together,
each prime number being raised to a power equal to
the Gödel number of the corresponding element in

the formula. The first ten prime numbers in order are
2, 3, 5, 7, 11, 13, 17, 19, 23, and 29, so the final Gödel
number for the foregoing formula is

$$(\exists x)(x = \mathbf{s}y) \longrightarrow 2^8 \times 3^4 \times 5^{11} \times 7^9 \times 11^8 \times 13^{11} \times 17^5 \\ \times 19^7 \times 23^{13} \times 29^9$$

We'll gladly leave it to the reader to compute the actual
value of this quantity! Using this kind of numbering
scheme, Gödel was able to attach a unique number to
each and every statement and sequence of statements
about arithmetic that could be expressed in the logical
language of *Principia Mathematica*.

It's not hard to see that this Gödel numbering
scheme doesn't differ much in spirit from the ASCII
coding procedure commonly used in computer science.
In that scheme, each of the various characters of the
English alphabet and punctuation symbols is coded by
a string of seven binary digits. Here are a few examples
of that coding

$$A = 1000001 \qquad 3 = 0110011 \qquad x = 1111000$$

Its main difference from Gödel's coding procedure
is that the ASCII scheme is designed to code only at
the single level of the individual alphanumeric symbols
used in the English language. On the other hand, the
Gödel scheme accounts for several levels of expression,
allowing us to distinguish between the lowest level of
a numerical variable, which is basically just the level
of the ASCII code, and the higher levels of sentential
and predicate variables that represent entire strings or
even properties of the lower-level variables, including
the all-important proof relations between axioms and
theorems. A key part of Gödel's route to incompleteness was to show how to code arithmetically these very
distinct semantic levels of logical expression.

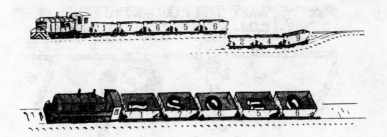

Figure 3.1. Freight train view of Gödel
numbering and transformation rules.

By Gödel's coding procedure, every possible proposition about the natural numbers can itself be expressed as a number, and this opens up the possibility of using arithmetic to examine its own truths. The overall process can be envisioned by appealing to the metaphor of a locomotive shunting boxcars back and forth in a freight yard. This idea, which originated with Douglas Hofstadter, is shown in Figure 3.1. In the upper part of the figure we see the boxcars with their uninterpreted numbers painted on the sides of the cars; below, looking down from the bird's-eye view, we see the interpreted symbols inside the cars. The shuffling of the cars in the switching yard in accordance with the rules for manipulating logical symbols and formulas is mirrored by a corresponding transformation of natural numbers—that is, statements of arithmetic—and vice versa.

Deep insights and profound results necessarily involve seeing the connection of several ideas at once. In the proof of Gödel's Theorem there are two crucial notions that Gödel had to juggle simultaneously. Gödel numbering was the first. Now let's look at the second Big Idea.

Figure 3.2. A self-contradictory self-referential statement.

Logical paradoxes of the sort that worried Hilbert are all based on the notion of self-reference. A humorous illustration of a self-contradictory type of self-referential statement is shown in Figure 3.2.

The granddaddy of all such conundrums is the so-called Epimenides (or Liar's) Paradox, one version of which is

> **This sentence is false.**

What Gödel wanted to do was find a way to express such paradoxical self-referential statements within the framework of arithmetic. Gödel needed such a statement in order to display an exception to Hilbert's thesis that all true assertions should be provable in a formal system. However, a statement like the Epimenides Paradox involves the notion of truth, something that logician Alfred Tarski had already shown could not be captured within the confines of a formal system. Enter Gödel's Big Idea number 2.

Instead of dealing with the eternally slippery notion of truth, Gödel had the insight to replace "truth" by something that is formalizable: the notion of *provability*. Thus he translated the Epimenides Paradox into the Gödel sentence:

> **This statement is not provable.**

This sentence, of course, is a self-referential claim about a particular "statement," the statement mentioned in the sentence. However, by means of his numbering scheme, Gödel was able to code this assertion by a corresponding self-referential, metamathematical statement expressed in the language of arithmetic itself. Let's follow through the logical consequences of this mirroring.

If the statement is provable, then it's true; hence, what it says must be true and it's *not* provable. Thus the statement and its negation are both provable, which implies an inconsistency. On the other hand, if the statement is not provable, then what it asserts is true. In this case the statement is true but unprovable, which implies that the formal system is incomplete.

Gödel was able to show that for *any* consistent formal system powerful enough to allow us to express all statements of ordinary arithmetic, such a Gödel sentence must exist; consequently, the formalization must be incomplete. The bottom line then turns out to be that in *every* consistent formal system powerful enough to express all relationships among the whole numbers, there exists a statement that cannot be proved using the rules of the system. Nevertheless, that statement represents a true assertion about numbers—one that we can see is true by "jootsing," to use Douglas Hofstadter's colorful term for "jumping outside of the system." Almost as an aside, Gödel also showed how to construct an arithmetical statement A, which translates into the metamathematical claim "arithmetic is consistent." He then demonstrated that the statement A is not provable, which implies that the consistency of arithmetic cannot be established by using any formal system representing arithmetic itself. Putting all these notions together yields

> **Gödel's Theorem—Formal Logic Version**
> *For every consistent formalization of arithmetic, there exist arithmetic truths that are not provable within that formal system.*

Because the steps leading up to Gödel's startling conclusions are both logically tricky and intricately intertwined, let me summarize, in Table 3.2, the principal landmarks along the road.

An indicator of the degree to which Gödel's results were unexpected can be found in the reaction to his original announcement of the theorem at a philosophy-of-science symposium in Königsberg, Germany, on September 7, 1930. Ironically, Königsberg happened to be Hilbert's home town, which may partially account for the lukewarm reception given to Gödel's presentation of his results. In fact, the transcript of the discussions at the meeting gives no indication whatsoever of Gödel's remarks, and an article published later, summarizing the papers given at the meeting, makes no mention of Gödel whatsoever! Like many belief-shattering ideas, Gödel's appears to have been so unexpected and revolutionary that even the professionals didn't at first understand what he had accomplished. But one participant who did immediately see the implications of the work was John von Neumann, who cornered Gödel after his talk and pressed him for more details—a case

Gödel Numbering: Develop a coding scheme to translate every logical formula and proof sequence in *Principia Mathematica* into a "mirror-image" statement about the natural numbers.

Epimenides Paradox: Replace the notion of "truth" with that of "provability," thereby translating the Epimenides Paradox into the assertion "This statement is unprovable."

Gödel Sentence: Show that the sentence "This statement is unprovable" has an arithmetical counterpart, its Gödel sentence G, in every conceivable formalization of arithmetic.

Incompleteness: Prove that the Gödel sentence G must be true if the formal system is consistent.

No Escape Clause: Prove that even if additional axioms are added to form a new system in which G is provable, the new system with the additional axioms will have its own unprovable Gödel sentence.

Consistency: Construct an arithmetical statement asserting that "arithmetic is consistent." Show that this arithmetical statement is not provable, thus showing that arithmetic *as a formal system* is too weak to prove its own consistency.

Table 3.2. The main steps in Gödel's proof.

of genius recognizing genius, perhaps. Over the next few months Gödel spoke about his theorem on several occasions in Vienna, finally publishing his epochal paper "On Formally Undecidable Propositions of *Principia Mathematica* and Related Systems" in 1931. The rest is history.

Gödel's results are not just a curiosity confined to the rarefied realm of mathematical logic. In addition to the deep philosophical consequences of incompleteness and its ramifications for epistemology and the limits to human cognition, Gödel's work sheds crucial light on the problem of mechanization of thought and computation. But this is a topic better left to a later chapter, where we will explore these matters in detail. For now, let's step back from the mathematics and philosophy and take a little longer look at the social and economic times that Gödel grew up in and how they helped shape his views on philosophy and mathematics.

Chapter Four

Young Gödel

Kurt Gödel's grandfather, Gustav Handschuh, settled in Brno in the middle of the nineteenth century. He came from the Rheinland, where he had worked as a weaver. In Brno he worked as a manager in the textile firm of Schöller. His very traditional wife, Rosita Bartl, came from the German-speaking Moravian town of Iglau. The Handschuhs lived at Bäckergasse 9 (now Pekorska), on the second floor of a typical Biedermeier house, which had an interior courtyard and open passages where neighbors would meet in the evening to chat.

Their daughter, Marianne Handschuh (1879–1966), later to be Gödel's mother, also lived in this house with her sister and two brothers. Gödel's father Rudolf (1874–1929) was also born in Brno, and he lived with his aunt and foster mother Anna on the first floor of the same house as the Handschuhs.

The paternal side of Gödel's family came from Austria. Rudolf Gödel's parents were born in Vienna. Rudolf Gödel himself was sent to live with his aunt in Brno, in the aftermath of the suspected suicide of his father. The Handschuhs were close to Aunt Anna, often playing music or engaging in amateur theater together. As a result, Marianne Handschuh and Rudolf Gödel became acquainted very early.

Soon after their marriage, Rudolf and Marianne moved to an apartment at Gomperzgasse nr. 15 (now Bezrucova). Their first son, Rudolf, was born here in 1902. Later they moved back to their childhood street, Bäckergasse 5, next door to the house where the Handschuhs and Aunt Anna had lived. This is where the future mathematical genius Kurt Gödel was born on April 28, 1906.

Kurt's older brother, Rudolf Gödel, reports:

> Family life was harmonious. I got along very well with my brother, as did both of us with our parents. When he was around eight years old my brother had a bad case of rheumatism of the joints with a high fever; since then he became a hypochondriac and imagined he had a heart defect, which was never medically established.
>
> In general, Kurt Gödel was a happy but shy child. He was very sensitive, and was called "Herr Warum" (Mr. Why) because of his great curiosity.

Before the start of World War I, the Gödel family moved from Bäckergasse to their own large house on Spielberggasse 8a (now Pellicova). Like his grandfather, Gödel's father worked in the textile industry, becoming a shareholder and director of the large Redlich factory. In this capacity, Rudolf Gödel was able to provide a high standard of living for his family. For example, they owned one of the first Chryslers in the Austro-Hungarian monarchy, and their mansion had a spacious garden with a garden house on the southern slope of the hilly region of Spielberg.

By standing on the roof and using a telescope, the Gödel sons could see the stone ornamentation on the

Gothic cathedral of Sts. Peter and Paul in Brno. When the skies were clear, they could look as far as the borders of Lower Austria, and watch the trains leaving the station for Vienna. From the windows facing the other direction, it was possible to see the notorious prison on the Spielberg's peak.

When he was six years old, Kurt was enrolled in Brno's private Protestant school. From 1916 to 1924, he attended the Staatsrealgymnasium—the type of German secondary school that emphasized modern languages and natural sciences—on the Strassengasse (today Hybesova), where he was an industrious and very gifted pupil.

Kurt and Rudolf as children

Gödel's brother reports that his interests ranged from languages, including Latin, to history and mathematics. From the first grade of primary school to his graduation from the Gymnasium, young Kurt received only the highest grades. His mathematical and geometrical talents first became apparent when he was 14, and by the age of 16 he was already studying Kant, whom he regarded as a formative influence for his intellectual development. In this connection Rudolf Gödel says,

After the war we were in Marienbad quite often with my brother, and I remember that we once read Chamberlain's biography of Goethe together. At several points, he took a special interest in Goethe's theory of color, which also served as a source of his interest in the natural sciences. In any case, he preferred Newton's analysis of the color spectrum to Goethe's.

Kurt had an especially good relation with our mother, who often played his favorite melodies (light music) for him on the piano. When I took a walk with my parents, he often preferred staying at home and reading a book.

Curiously, Kurt Gödel would retain his love of light music throughout his life, preferring Strauss's waltzes to the music of Bach or Wagner, whom he in fact detested.

About Kurt's earlier years, Rudolf adds,

He could admire nature, but he did not really show a genuine love for it, as our mother did. Although my brother was perhaps less close to the family and went his own way earlier, he was also—particularly later on when he had become rather sickly—mother's special problem-child.

This latter observation refers to the bout of rheumatic fever that Gödel went through at the age of eight. Although the disease seems to have left no lasting damage, the affliction was a turning point in Gödel's life. His inquisitive nature caused him to read about the disease, and he thus learned about its possible side effects, including cardiac damage. Despite doctor's assurances, he came to believe that his heart had indeed been affected, an unshakable conviction that he carried for the rest of his life. His brother always claimed that this was the source of the hypochondria that was such an integral part of Gödel's later life.

Given their wealth and high social status, the Gödel family was able either to avoid or, when necessary, to overcome the difficult circumstances of daily life in this period of reconstruction after the ravages of World War I. Despite the political, social, and economic unrest in the still shaky nation–states of Czechoslovakia and the German part of the former monarchy, the family's tranquil life continued. And as long as their father was still alive, the two sons, Rudolf and Kurt, could pursue their studies with no particular financial concerns.

Rudolf Gödel had already moved to Vienna in 1920 to study medicine at the university. As he recalls,

> All my teachers were, in fact, world-famous. The internist was Wenckebach, the surgeon Eiselsberg, the pediatrician Birkee, the neurologist Wagner-Jauregg. Each was greater than the next.

When Kurt joined his brother four years later, in the autumn of 1924, he took an apartment in the eighth district of Vienna at Florianigasse 42, where he remained until April 1927.

At the university, young Gödel first studied mathematics, physics, and philosophy. He initially intended to specialize in theoretical physics, and he attended the lectures of Hans Thirring, which were held in the great lecture hall on the fourth floor of the Institute for Theoretical Physics, located in Vienna's ninth district at Strudelhofgasse 4. Serendipitously, the university's Institute for Mathematics was housed in the basement of the same building. And in 1926, two years after beginning his studies, Gödel decided to "move downstairs" and make mathematics the central focus of his studies.

It was not uncommon for Gödel to fill in for the major professors from time to time by giving a lecture or helping advise younger students in the department. Professor Edmund Hlawka of the Technical University of Vienna, who was one of Gödel's students at this time, looks back on Gödel's decision to abandon physics in favor of mathematics:

He was of course most strongly influenced by Hans Hahn and Karl Menger, attending their lectures on set theory and real functions. He also attended Furtwängler's lectures on number theory. And, I believe, that was what prompted him to apply number-theoretical methods to logic—to represent the propositions of logic and mathematics by natural numbers, what is now termed "Gödel-ization."

Gödel's teachers at the institute included Hans Hahn (one of the founding members of the Vienna Circle), Karl Menger (son of the distinguished economist Carl Menger), Phillip Furtwängler (cousin of the noted conductor). Gödel's most important professors of philosophy were Moritz Schlick (the leader of the Vienna Circle), Heinrich Gomperz (son of the distinguished classical philologist), and later the philosopher of science Rudolf Carnap.

Gödel and the Vienna Circle

Many of Gödel's teachers were members of the Vienna Circle, the famed group of philosophers, mathematicians, and scientists who met in a seminar room at the Mathematics Institute every Thursday evening to develop a theory of scientific "truth." Thus Gödel had already developed his links with the people and ideas of this influential group during his student days, attending their meetings regularly from 1926.

The meetings were by invitation only, so presumably Gödel was invited by Hahn or Schlick. The first time Gödel attended was when the group was reading Wittgenstein's *Tractatus* (for the second time!), and he continued to attend regularly until 1928. But Gödel went to some pains to make it clear that he did not agree with the Circle's views. In particular, he disagreed with the idea that mathematics should be regarded as the "syntax of language," a position maintained especially by Carnap. While holding pretty strong views on this and other matters, Gödel also tended to avoid controversy, and so he did not actively criticize the positions of other members of the Circle. Mostly he was content to listen to what others had to say, only occasionally injecting his own ideas into the discussion. It seems likely that the expression of contrary views helped stimulate Gödel to develop his own ideas more clearly. But in retrospect, it appears that the main impact the Vienna Circle had on Gödel was in introducing him to new literature and bringing him into contact with colleagues that he would not have met otherwise.

From October 1929 on, Gödel was a frequent visitor to Karl Menger's colloquia, which took place at the Mathematics Institute. There Gödel met many leading figures of the mathematical world, such as the Polish logician Alfred Tarski, the Hungarian polymath John von Neumann, and the German statistician Abraham Wald. Karl Menger was the editor of the *Results of a Mathematical Colloquium,* and he soon made Gödel a co-editor, along with Georg Nöbeling and Franz Alt. In this period, Gödel was especially prolific, publishing 13 mathematical articles between 1929 and 1937.

Among those scholars who regularly participated in the colloquia, Gödel had the reputation of being extraordinarily gifted. His company was eagerly sought after—especially, it seems, by women. The number theorist Olga Taussky-Todd, a student with Gödel, recalls one incident in particular. Todd was working one day in the small seminar room outside the library in the Mathematics Institute, when a slight, young woman entered the room. The woman set down her books and began to work. She was quite good-looking, with a slightly gloomy face, and wore a beautiful, rather unusual summer dress. Not much later Gödel walked into the room.

The two of them began to talk—Gödel all charm—and they soon left together. Todd eventually befriended the girl, who intimated that she and Gödel had indeed had an affair (she complained that Kurt was spoiled and had to sleep too long in the morning.) Evidently, somewhere between Brno and Vienna, Gödel lost his reclusive shyness and entered into quite a social whirl during his student days in Vienna.

In the short period between October 1927 and June 1928, Gödel lived at Währingerstrasse 33. The Café Josefinum was located on the ground floor of this house, where Gödel was a frequent visitor, often with his colleagues from the Vienna Circle. In July 1928 Gödel moved to a spacious apartment at Langegasse 72, where he stayed until November 1929. This flat was actually intended for use by his parents, should they ever decide to visit Vienna. This was a fateful move for Gödel, for directly across the street lived Adele Nimbursky, who was then married to a photographer but who was later to become Gödel's wife. She performed in the cabaret club *Nachtfalter,* located in the center of the city on Petersplatz. Although there is no documentary record of their first meeting, Gödel probably met the attractive Adele on Langegasse during the time when he was working on his doctoral dissertation.

The courtship was a very protracted one, because Gödel's parents disapproved strongly of Adele. In their eyes she was unsuitable for Kurt for a variety of reasons: She was divorced, was older than Kurt by six years, was Catholic, came from a lower-class family, and was a *dancer.* She herself claimed to be a ballet dancer, though if this is true, she must have been at best an extra in the corps de ballet at the Volksoper. In any case, the social stigma was great; in those times, a ballet dancer was hardly more than a high-class prostitute. To marry

Marriage photo of Kurt and Adele

such a person could destroy even a well-established career.

Gödel's father died on February 23, 1929, at the age of 54. Soon after, his widow moved to Vienna. As Rudolf recalls,

> [With this death] our world collapsed, especially for our mother. For months on end, her condition led us to fear the worst. It was impossible for us to leave her alone in Brünn. The villa with its garden was rented out, and our mother, my brother, and I moved into a large apartment in the Josefstadt (the doctors' district of Vienna).

The district was originally one of Vienna's suburbs. It was named in honor of the coronation of Kaiser Josef I. Here nobility, clergy, and the prosperous Viennese bourgeoisie built their palaces, churches, and splendid homes. The Gödel family lived at Josefstädterstrasse 43 until November 1937. Among Gödel's many residences, this is the one best known by the scientific community. Between 1929 and 1937 he wrote his most important articles here, while carrying on correspondence with mathematicians from all over the world—Ernst Zermelo in Freiburg (Germany), Jacques Herbrand in Paris, Oswald Veblen in Princeton, Paul Bernays in Zürich—as well as with John von Neumann,

the most cosmopolitan of them all, who lived simultaneously, it seemed, in Budapest, Zürich, Berlin, Princeton, and Los Alamos.

During this same period, Gödel's mother encour-
aged the brothers to take an active part in Vienna's
cultural life, and she frequently accompanied her two
sons when they attended performances at the Josef-
stadt's theater. Books and music were other diversions.
Along with Goethe and Kafka, Gödel's favorite authors
included the moody novelist Stefan Zweig, Arthur
Schnitzler (whose novella *The Dream Novel* was the basis
for Stanley Kubrick's film *Eyes Wide Shut*), and the Austrian social commentator and novelist Heimito von
Doderer. In the musical realm, Gödel loved waltzes,
Italian opera, the operettas of Johann Strauss, and the
operas of Richard Strauss. In accordance with his general musical taste, Gödel preferred the lighter music
of his time, which included both Jacques Offenbach's
Barcarola and Franz Schubert's songs, as well as American hits such as "Harbor Lights" and "The Wheel of
Fortune." Gödel also cultivated an admiration for the
lovely singer Maria Ceborati.

But it was really the Vienna Circle that provided
Gödel's intellectual sustenance. The group met every
Thursday at 6 P.M. in the Mathematics Institute on Strudelhofgasse in Vienna's ninth district, from 1924 to
1933. Hans Hahn (1879–1934), professor of mathematics at the University of Vienna and Gödel's teacher, was
the Circle's founder. In 1922 he arranged for Moritz
Schlick to be appointed to the faculty of the university. He also made it possible for Schlick's assistant,
Karl Menger (1902–1985), to teach there as an associate professor from 1927 on. Together with Philipp
Frank (1884–1966), Hahn's friend from student days,
and Otto Neurath (1882–1945), this group of thinkers

realized their dream of founding a creative base for scientific progress in Vienna.

Otto Neurath was the organizing force behind the Vienna Circle. Neurath was also the leader of the circle's liberal wing, which sought to promote social change and met regularly in the Ottakring center for adult education and culture. Along with many others, Neurath was forced to flee Vienna's Austro-Fascists under the Dollfuss government in 1934.

Another of Gödel's teachers—and later colleague—was Rudolf Carnap (1891–1970). He was the most radical of the Vienna Circle's philosophers, espousing the belief that the problems of philosophy are all reducible to the problems of language. Carnap became the Circle's most visible member following his emigration to the United States in 1935.

Other leading members of the Circle included mathematician Herbert Feigl, set theorist Karl Menger, philosopher Viktor Kraft, legal theorist Felix Kaufmann, mathematician Friedrich Waismann, and philosopher Edgar Zilsel. Among the many distinguished visiting thinkers from abroad who attended these meetings were the philosophers Alfred Ayer, Carl Hempel, and Willard van Orman Quine.

In addition, loose discussion groups formed at the Circle's periphery around historian Heinrich Gomperz, physicist Richard von Mises, and philosophers Karl Popper and Ludwig Wittgenstein. There was regular intellectual interchange between these groups and the Vienna Circle itself, the central core of this modern, innovative movement.

As we noted earlier, Gödel became a member of the Vienna Circle in 1926, at the very outset of his mathematical studies. Both in the Circle meetings and in the colloquia organized by Karl Menger, revolutionary ideas in logic often came up during the course of debates. Of special interest were the discussions centering on the work of John von Neumann (1903–1957) and Alfred Tarski (1901–1983). One of these meetings served as the venue for Gödel's first research report.

The decades-old debate on the interrelationship of language, world, and science revolved around one central problem: the status of our perceptions and our ability to express them. A new form of mathematical logic had been ushered in by Frege's formalization; by Mach's critique of inherited scientific language; by Fritz Mauthner's nominalistic critique of language and society; and, in particular, by Bertrand Russell's elaboration of the ideas of Peano in the ground-breaking book *Principia Mathematica,* wherein he and Alfred North Whitehead laid down the logical foundations of arithmetic. This logic led to more precise formulations and basic claims aimed at establishing a new scientific philosophy designed to convey a new perspective on the way science and the world are linked through logic. According to the members of the Vienna Circle, traditional philosophy produced nothing more than "word music" and "conceptual poetry"—a mode of thinking confined now to the world of art.

A pressing issue for the Vienna Circle was to eradicate metaphysics. Hence, metaphysical propositions had to be clearly distinguished from propositions that were truly scientific. To this end, Schlick strongly advocated the notion of verifiability. His assistant, Friedrich Waismann, concurred: "If you cannot specify any way of knowing that a given proposition is true, the proposition has no meaning whatsoever; for the meaning of a proposition resides in its method of verification." In addition, the Circle saw its task as creating a unification of various realms of cognition. The basic science and role model for this was taken to be physics. Because applying this reductionist view met with some success in areas such as chemistry and biology, Neurath and Carnap advanced the amazing claim that every proposition must be translatable into the language of physics.

Ludwig Boltzmann's elevation of statistical probability to a general method of analysis within physics made it possible to consider a large number of different factors. As he stated, "Statistics should allow us to consider the total influence of the infinite universe upon the nearly isolated partial system." Boltzmann's formula for his Entropy Principle was one of the first applications of the statistical method. The impetus for the development of mathematical logic thus emerged not only from mathematics but also from evolving ideas in modern physics. (Recall Gödel's early study of theoretical physics and his later formulation of a new solution to Einstein's field equations in relativity.)

As an advocate of atomism, the notion that all matter is composed of elementary particles called atoms, Ludwig Boltzmann made a deep impression on the Vienna Circle. This was due, in particular, to his concept of what constituted a scientific model. For Boltzmann, science does not grasp the reality of nature itself but, rather, offers models of nature. The solar system, for instance, serves as a model for the structure of atoms. Such models change according to our theories, and they need to be both logically free of contradictions and subject to empirical scrutiny (that is, they must agree with the experimental data). They also must contain as much information as possible and must be compact (that is, minimally redundant).

The precise formulation of a theory of models was closely connected with Boltzmann's method. It has played a prominent role in quantum physics and is important today in the conception of data banks and other software products within computer science.

With its proposition of a mathematical-logical notion of language, the early Wittgenstein's *Tractatus Logico-Philosophicus,* along with his approach to mathematics in general and his debates with Bertrand Russell, was discussed intensely in meetings of the Vienna Circle. (Interestingly, Russell's ideas were themselves a starting point for Gödel's research.)

Wittgenstein illustrated these ideas by what he called a "picture theory" of language, in which he compared logical propositions to pictures. A picture can represent some physical state via certain types of symbols; language can do likewise, but with a different set of symbols. The picture bears some relationship to the physical reality that it represents. Thus, for example, if we see a human face in a photograph, the nose may appear in the center of the face both in physical reality and in the picture. However, if the picture is by Salvador Dali, the nose might appear in some quite different location, or not at all. Of course we might try to clarify the relationship between the picture and the object—for example, by introducing color or perspective—but such an attempt at clarification only gives rise to another picture, which itself will require additional analysis. At some stage the essence of the picture has to be understood directly, or we fall into an infinite regress.

In the picture theory of language, propositions making up the language are thought of as analogous to a series of pictures. Furthermore, because Wittgenstein assumes that the logical structure of language mirrors the logical structure of reality, the language "pictures" represent *possible* states of the world. It follows that linguistic statements are meaningful when they can, in principle, be correlated with the world. Actual observation of the world will then tell whether they are true or false. To illustrate, we can meaningfully say that "the United Nations is in New York," but it is meaningless to state that "is United the New in York Nations." Of course, different logical rules (grammars) could be developed in which the latter statement would be meaningful, but within the context of conventional English grammar it has no logical structure at all. Thus the main claim of the picture theory—that there must be something in common between the logical structure of the language and the structure of the fact that it asserts—cannot really be "said" in terms of the language being used to make the statement; it can only be "shown." This conclusion gave rise to Wittgenstein's famous metaphor in the penultimate section of the *Tractatus:*

> My propositions serve as elucidations in the following way: anyone who understands me eventually recognizes them as nonsensical, when he has used them—as steps—to climb up beyond them. (He must, so to speak, throw away the ladder after he has climbed up it.) He must transcend these propositions, and then he will see the world aright.

Wittgenstein's punch line, then, is that the sense of the relationship between reality and its description in language cannot be expressed in language.

Thus ended Wittgenstein's "early period" studies on the interplay of logic, language, and reality. The essence of his ideas can be summarized in the following steps:

1. There is a world that we want to describe.

2. We try to describe it in some language—scientific, mathematical or otherwise.

3. There is a problem about whether what we say about the world corresponds to the way the world really is.

4. We want to know the true nature of the correspondence between what we say and the way things are, but we can use only language itself to describe that correspondence.

5. Words of a language can never express the desired correspondence, and we must take recourse merely to *showing* it—that is, using the picture theory—because otherwise, we would fall into the infinite regress of descriptions of descriptions of descriptions . . . .

At Step 5 we come to one of the most famous statements in all of philosophy, with which Wittgenstein concluded the *Tractatus:* "That whereof we cannot speak, we must pass over in silence."

Wittgenstein's basic concept in the *Tractatus,* a one-to-one correspondence between pictures of the world in the form of propositions and actual states of the world itself, bears a striking resemblance to the way computers function. This idea influenced John von Neumann, who would later, in the 1940s, develop the idea of a stored program to control the operations of a computer. Gödel himself never actually met Wittgenstein, although he did see him once at a lecture given by L.E.J. Brouwer on mathematics, science, and language at the University of Vienna in 1928. Considerable distance would in any case emerge between Gödel's perspective and Wittgenstein's later concept of mathematics and language as a "game" and a reflection of a form of life.

As a result of general theoretical differences over the way one verifies the truth of a proposition in lan-

guage, Wittgenstein eventually distanced himself from
the Vienna Circle, especially because its manifesto of
1929 struck him as being excessively polemical. In his
posthumously published *Logical Investigations,* Wittgenstein would alter his mathematical–analytic approach
to language, favoring a far more socioculturally ori-
ented position.

By way of contrast, Gödel based his ideas on anything but sociocultural assumptions. Moreover, he distanced himself from Carnap's formal–syntactic interpretation of science and mathematics. Throughout his
life Gödel was a self-professed "realist" and "Platonist";
he even attempted an ontological proof of the existence of God.

The limits and potential of language, logic, and
meta-language also played a central role in the critical social theory of Fritz Mauthner (1849–1923), who
also analyzed language and logic, though in a rather
idiosyncratic way. Nevertheless, these ideas influenced
Wittgenstein, as indicated by the fact that Mauthner was
one of the few people explicitly mentioned in the Pref-
ace to Wittgenstein's *Tractatus.* Realism ascribes existence to general concepts. Mauthner was critical of this,
seeing this point of view as one of the main foundations
of the abuse of power. But he also saw such "linguistic
superstition" as leading to a mistaken approach to scientific analysis. Mauthner's position was thus similar to
that of Mach. In a letter to Mach, he indicated that
"the book [*Contributions to a Critique of Language, I–III*]
will have an especially revolutionary impact on people
standing far closer than myself to the field of linguistics,
awakening them from a 'dogmatic slumber.'"

Wittgenstein applied his own principle that "the
boundaries of my language signify the boundaries of
my world" to mathematics. In other words, the objects

of mathematics were understood as being limited to those entities that could be formulated in mathematical language. It is implicitly assumed here that mathematical truths can in fact be arrived at through formalized mathematical language.

This mathematical language is the language of predicate calculus, including its mechanisms of proof. Such a calculus emerged from the schematization of language into linguistics—the construction of grammatical rules and syntactic forms. In this sense, a parallel exists between the goals and accomplishments of mathematical and of scientific theory.

David Hilbert (1862–1943) attempted to formalize mathematics with the help of predicate calculus. Under his leadership in Göttingen, then the center of mathematical research in Germany, language was regarded as a formal tool for furnishing proofs. By way of contrast, the Vienna Circle treated language as a philosophical subject, one approached critically from a theoretical perspective.

In numerous discussions inside and outside the Circle over several decades, Gödel thus came into contact with a rich tradition from which he took much direct intellectual inspiration. It would lead him away from a formal concept of language and toward an intuitive perspective that would interact with a form of language-critical realism. In any event, it seems that this development was congenial to Gödel's personality: He first described himself as a mathematical realist in 1925, more than a year before he abandoned physics and began his formal mathematical studies.

It also seems clear that if Gödel had not studied in Vienna, he would never have plumbed the depths of mathematics to discover his "principle of incompleteness" and prove it mathematically. This brilliant discovery refuted both Hilbert's program for mathematics and Wittgenstein's equating of the boundaries of language with the boundaries of the world. Later, in his second philosophical phase, Wittgenstein himself would come to consider Gödel's discovery as a new approach to solving the problem of provability. In contrast to Hilbert and Wittgenstein, Gödel saw that languages also have their limits; using Hilbert's method, he demonstrated the limits of formal systems.

Gödel thus showed us that the mathematical world is more complex (and hence stronger) than mathematical language. Language is itself sometimes more precise than thinking, but it is simultaneously weaker in that its syntax does not allow a reconstruction of all conceivable models.

What can be proved within and through language is less than the capacity of human thought. This, in turn, is less (weaker) than what is possible in the world.

A tour de force like Gödel's could have been possible only during a period of extreme intellectual concentration—and the Vienna of the 1920s and 1930s provided exactly this. Gödel mathematically unfolded the labyrinth of self-reference. His work illuminated the path into and out of this labyrinth by means of rational rules.

Chapter Five

Life in Princeton

Within Austrian society, the collapse of the First Republic and the installation of a fascist regime in Austria in 1933 marked the end of efforts at modernization. Virtually all members of the Vienna Circle were sympathetic to the Social Democratic Party, which was then outlawed. The large-scale persecution of liberal and leftist social forces sparked a general exodus of these philosophers, scientists, and mathematicians, a process that culminated in the *Anschluss*—the "annexation" of Austria by the German Reich.

Gödel left Vienna for Princeton in January 1940. The Austrian political upheavals had gradually revealed their true face over the previous decade, ushering in the demise of cultural and scientific life. For a long time, the splendor of the First Republic had been a mere façade, displaying an architecture of beautiful illusion, but riven by internal contradictions at the core. The labor unions had been banned along with the Social Democratic Party in 1934; moreover, social progressives were continuously prosecuted. These measures, together with others such as new restraints on press freedom, severed the ties between progressive ideas and social life, causing the culture to wither away.

From a modern perspective, both Austrian fascism and German Nazism appear to have mounted a crusade against the new world of rational science. The connection between the ends and means of science was broken, allowing a kind of paranoid structure of governance to emerge. Already in 1928, six years before the fascist takeover, most Austrian students identified with German nationalism, a tendency reflected in increasing efforts to break up seminars and lectures held by Jewish, socialist, and liberal, leftist professors. The attacks—particularly those against the Vienna Circle—became increasingly violent. They culminated in the murder of Moritz Schlick, who had been systematically decried as a Jew (in fact, his family stemmed from old Austrian nobility). Schlick was gunned down on June 22, 1936, on the steps of the University of Vienna.

The murderer was Schlick's former student Hans Nelböck, who was inspired by both the general public mood and the views of fellow student Leo Gabriel, who was himself a Catholic Austro-Fascist. After World War II, Gabriel would become professor of philosophy at the University of Vienna. Nelböck was released after two years in prison, following the triumphant entry of Hitler's troops into Vienna in March 1938.

For years, "blacklists" had circulated among the Nazi students, identifying suspect intellectuals and academics. These included Erich Heintel, who would himself become a professor of philosophy at the university of Vienna after his speedy postwar "denazification." The blacklists focused on Jewish teachers and their close friends and colleagues. Despite the fact that he was not Jewish, Gödel's name was on one such list, because he had been a student of the "half-Jew" Hans

Hahn and was a member of the "Jewish" Vienna Circle. In the same demonizing spirit, fields such as mathematical logic and set theory, along with Einstein's relativity theory, were denounced as "Jewish." This was the basis for an assault on Gödel perpetrated near the Mathematical Institute by radical right-wing students in early November 1939. (The reason for this attack is unclear. It is probably attributable simply to the fact that there were a lot of young punks cruising the streets of Vienna during this period, randomly mugging people who looked Jewish, intellectual, or both.) Luckily, Adele, his wife-to-be, had her umbrella with her that day, and managed to fend off the attack before harm was done to either of them. It's likely that this episode brought home even to someone as otherworldly as Gödel, that Vienna was a dangerous place—and getting more dangerous with each passing day. No doubt a life in the tranquil groves of academe on the other side of the Atlantic looked a lot more attractive to Gödel than ever before.

Gödel always rejected totalitarian ideologies. (As we shall see in a moment, he would even come to fear for democracy in his new American homeland.) But in contrast to Einstein, for example, he never took a public political stance, and his intense preoccupation with mathematical issues suggests that he had no wish to experience the political changes around him directly. Nevertheless, these changes inevitably had their effect on both his personal and his intellectual life. The persecution of liberals and leftists meant that many of his best friends and teachers were forced to emigrate.

In addition to Schlick's murder in 1936, we should also note the emigration to America of his mentor and interlocutor, Karl Menger, the following year—a choice made by famed mathematician John von Neumann,

Oskar Morgenstern, and statistician Abraham Wald as well. Likewise, with the help of Karl Popper, Friedrich Waismann moved to England in 1938. And Wittgenstein himself had been based in Cambridge since 1929. Gödel's brother Rudolf speaks of this period:

> The year 1933—the year of the Nazi rise to power in Germany—does not evoke any special memories for me in Vienna. Since none of us was very interested in politics, we could not properly assess the significance of the event.
>
> Two events then quickly opened our eyes: the murders of Chancellor Dollfuss and the philosopher Schlick, in whose circle my brother had been active. The latter event certainly triggered a severe nervous crisis in my brother, so that he had to spend some time in a sanitarium—which was naturally a source of great concern to my mother. Soon after his recovery, my brother was invited to take up a guest professorship in the USA. This and the fact that life was becoming increasingly expensive for us in Vienna, since our funds were in Czechoslovakia, prompted our mother moving back to our villa in Brno—it was 1937, in other words a year before Hitler occupied Austria. There were difficulties with the Villa's Czech housekeeper, since the Czechs naturally felt a deep hatred toward the Germans then, on account of the repeated executions of Czech "traitors"—these publicly posted on red posters. This is how our mother spent World War II in Brno.

Thus the political situation separated the Gödel family. For his part, Rudolf would remain in Vienna, serving as director of a large radiography institute, a position that protected him from the military draft.

Between 1933 and 1940, Gödel shuttled back and forth between Vienna and the newly founded Institute for Advanced Study, still located then in Princeton University's Fine Hall. His first visit lasted from autumn of 1933 to May 1934. After returning to Vienna for a year, he went back to Princeton in 1935. But he had to return almost immediately to Vienna for reasons of health.

In 1938 Gödel accepted an invitation to fill a temporary position at the Institute, which ultimately led to his
final position there as professor. On account of all this
commuting between Princeton and Vienna, Gödel was
often forced to cancel his lectures at the Mathematical
Institute in Vienna. But by the same token, his irregu-
lar absences spared him the chaos of the civil strife in
Austria in February 1934.

After completing his *Habilitation* under the direction of Hans Hahn (with Wilhelm Wirtinger, Hans Thirring, and Karl Menger as referees) in 1933, Gödel became a *Privatdocent* at the University of Vienna. In this
capacity he had the *Venia Legendi*—the right to give
lectures. He did not have any regular income, however, and was thus forced to live on the inheritance
he received from his father. Despite his groundbreaking accomplishments in mathematical logic, the University of Vienna did not offer Gödel a position com-
mensurate with his achievements; this slight was clearly
a reflection of the political context of the time. Immediate recognition of Gödel's brilliant work by the Institute for Advanced Study thus partly rectified an unjust
situation.

Gödel's lectures at the Institute for Advanced Study
contributed substantially to the development of an
American school of logic founded by Emil Post and
Alonzo Church. Gödel was the first to define more precisely the concept of an algorithm, using the concept of
a recursive function—essentially, a function for which
there is a mechanical rule for computing the values
of the function from previous values, one after the
other, starting with some initial value. This step formed
the basis for the *Principle of Computability,* which is now
indispensable to theoretical work in computer science.
It enabled Stephen Kleene to develop his theory of

(partially) recursive functions, which made the notion of an algorithm even more precise. Church, in turn, advanced what is now called Church's Thesis, which asserts that the precisely defined notion of a recursive function can be identified with the intuitive, informal concept of what it means to carry out a "computation." And in the same context, a student of Gödel's, J. Barkley Rosser, made the *Principle of Incompleteness* even more precise.

On the personal front, the most decisive event in Gödel's life was his marriage to Adele Nimbursky on September 20, 1938. Several years older than Gödel and previously married to the photographer Nimbursky, Adele (née Porkert) had a practical bent and a very optimistic nature. She acted as a veritable life-line to Gödel, bestowing on him a level of attentiveness that was both maternal and intense. Early in their time together, Adele defended Kurt from the Nazi Brownshirts, fending them off with her umbrella, and later she served as a "taster" of his food when his paranoia convinced him that someone was trying to poison him. In these periods, Adele fed him spoonful by spoonful, until she was able to bring his weight up from 106 pounds to a more respectable 140 pounds.

Just two weeks after the marriage, Gödel said goodbye once more to both Vienna and his wife in order to work again at the Institute during the 1938/1939 academic year. Meanwhile, the new German authorities had withdrawn his status as a *Privatdocent* at the university at the request of the head of the philosophy department, a certain Professor von Christian. This request had been approved by the head of the Docent Association, Dr. Marchert. Surprisingly, Gödel applied for a new appointment as a "Docent of the New Order" to replace his earlier *Privatdocent* position. And he actually received this position a year later, after he had already settled in Princeton, when the application was approved by the Ministry of the Reich in Berlin.

Following his final return to Vienna in 1939, Gödel was called up to serve in the army, having been found, despite his always precarious health, fully qualified for front-line service. He immediately turned to Oswald Veblen, director of the Institute for Advanced Study, who again offered him a visiting position. Gödel and his wife were thus able to leave Austria in January 1940. Unable to cross the Atlantic, however, because of the British blockade, the couple first went to Japan via the Trans-Siberian railway and then sailed from there to San Francisco. They arrived in Princeton in March 1940. Gödel would never again set foot on European soil.

The Institute for Advanced Study

Princeton was Gödel's home for the second half of his life. It was there that Gödel further developed many aspects of his mathematical work, while concerning himself increasingly with the philosophical implications of his mathematical discoveries.

Now a rather congested suburb, Gödel's Princeton was still a small, green New Jersey town, then as now an hour's drive from New York City. Its quiet, wood-framed, colonial houses bordering the university's lush, neo-Gothic campus based on that at Cambridge, UK, created an atmosphere ideal for a mathematician like Gödel. The goal of the Institute for Advanced Study was to furnish superb working conditions for the most distinguished scientists and scholars. There were no teaching duties, no students, and very few committees or other distractions from the contemplative life of the scholar.

Along with Einstein, Gödel found himself in the company of the famed physicist and director of the Manhattan Project, J. Robert Oppenheimer, as well as world-renowned mathematicians John von Neumann, Hermann Weyl, Oswald Veblen, and many other legendary figures. Between 1940 and 1946, Gödel's appointment at the Institute was temporary, requiring annual renewal, but in 1946 he was granted a permanent position, and in 1953, at Einstein's and von Neumann's urging, he was finally appointed full professor.

Given that Gödel's groundbreaking work in mathematical logic had been done 20 years earlier, it's odd that it took so long for the IAS faculty to promote him to this level. One possibility is that there were "opponents" to Gödel's promotion, people who felt there were already enough "crazy" men on the Institute's faculty. Others attribute this unseemly delay to Gödel's overly legalistic turn of mind, believing that his lengthy deliberations on matters such as the invitation of visitors to the Institute would retard the orderly progress of business at the IAS. One account of his administrative activities at the Institute is quite revealing in this regard. Deane Montgomery, a fellow faculty member, recalls that

> He was actually a highly conscientious faculty member, participating enthusiastically in faculty life—very much in contrast to what had been anticipated by some people . . . . He was especially interested in the Institute's appointments, both the temporary ones and, in particular, those which were long-term. Gödel's difficulties in determining the qualifications of various candidates or what qualifications were required for membership appeared to stretch out the faculty meetings. So we decided to change things—to form a separate committee for logic. I myself volunteered for this committee. Gödel and I thus established regular contact, mostly on the telephone. For he preferred simply speaking to me about

certain things on the phone, rather than meeting personally, even when we were both directly present in the Institute . . . . And when invited to come by to discuss things directly he would say, 'Just tell me about it on the phone. We can arrange things on the phone.' And sometimes I would say 'I have the feeling this subject is too complex. I can't discuss this on the phone.' He would then allow me to come by and discuss things directly. He seems to have felt more secure on the telephone, maybe because he could hang up if he felt too tired. I don't know why. Possibly he found anyone who came too close a burden, and that caused him difficulties.

Gödel preferred the telephone for private communication as well—he would often speak for hours with people all over the continent. In general, he had the reputation of being withdrawn, and he cultivated limited but intensive contacts, the most notable being his daily walk to and from the Institute with Einstein. He also avoided public appearances; his last talk was given in 1951 as the Gibbs Lecture to the American Mathematical Society. Responding to a symposium in Ohio in honor of his 60th birthday, he simply sent a telegram. Among his colleagues at the Institute, his only close friendships were with Einstein, the economist Oskar Morgenstern, and the logician Abraham Robinson.

Gödel considerably cut back his direct work on problems in mathematical logic at Princeton, perhaps because the solution to certain problems, such as those of set theory, proved very elusive. He did continue work on the problem of the Axiom of Choice and on the Continuum Hypothesis. The first is the seemingly simple assertion that given a collection of sets $A, B, C, \ldots$, it is possible to form a new set by "choosing" an element from each of this (possibly infinite) sequence of sets. The Continuum Hypothesis is closely related. It simply asserts that there is no level of infinity between the countable infinity of the natural numbers 1, 2, 3, ... and the uncountable infinity of the real numbers. But compared with the productivity of his Viennese period, one notices a marked difference.

Gödel's seeming decline in mathematical productivity during his years in Princeton is probably due to a constellation of factors. First of all, it's quite natural for mathematicians to do their most innovative work when they are young and to spend their later years polishing their earlier work and even philosophizing about it. In Gödel's case this is exactly what happened. In particular, his intellectual interests took a decided turn toward philosophy, including questions about the existence of God and the transmigration of souls. Other contributing factors to the fall off in his mathematical output were undoubtedly his increasing concern for his health, and his withdrawal from nearly all social contacts with other mathematicians. Given all these considerations, it would be surprising if Gödel had been able to continue to produce world-class mathematical results. And he didn't.

Gödel's foundational work in logic having already entered the mathematical canon by the time of his appearance in Princeton, in the early 1940s Gödel began to concentrate on the philosophy of mathematics. His philosophical interests had already become apparent during his youth, while he was in the Gymnasium, and had continued throughout his years in Vienna. His grappling with concepts under discussion in the Vienna Circle—particularly those put forth by Rudolf Carnap—need to be seen against this backdrop. Above all, Gödel was interested in the structure and analysis of axioms for formal logical systems, as well as in their context (space, time), universal concepts applicable to any set, and what one means by a "proof." Strangely, perhaps, in the 1950s, Gödel became interested in the notion of the transmigration of souls.

Essentially, Gödel held that an afterlife must exist because the universe basically has a meaning. His argument was that human potential is never fulfilled in a single human lifetime. Thus there must be an afterlife in which this potential could be fulfilled; otherwise, human life would be meaningless. This argument is reminiscent of various anthropic-based explanations of why the universe looks as it does and not like something else. Interestingly, Gödel's belief in an afterlife had nothing to do with a belief in God. In fact, there is no evidence that he believed in a Christian-type deity. His arguments were, as always, rationally based on the principle that the world and everything in it has meaning, or reasons. This is closely related to the causality principle that underlies all of science: Everything has a cause, and events don't just "happen."

During the 1940s and 1950s Gödel published a portion of his work on Russell (1942), Einstein (1949), and Carnap (1953); today, these are basic papers in the philosophy of mathematics. Other works, however—on Leibniz (1943–1946), Kant (1947), and Husserl (1959) —were never released for publication and surfaced only after Gödel's death. Seeing this uneven achievement as a reflection of a lack of fruitful dialog with his peers, Gödel's mentor, Karl Menger, maintained that Gödel's genius had really been wasted at Princeton.

In his philosophical work, Gödel took a Platonist point of view. Hao Wang described this position in the following way:

> Philosophy as precise theory can be considered an application of Gödel's conceptual realism. It is meant to make possible the correct perspective, illuminating the clarity of the underlying metaphysical concepts. Put somewhat

> more clearly, he states that the purpose of this exact the-
> ory is to determine the fundamental concept C and to
> find its axiom A, so that C alone fulfills the axiom, and
> A is part of our original intuition of C . . . . This ideal is
> closely connected with other aspects of Gödel's philoso-
> phy. For instance, he explains that in its general contours
> it corresponds to the (metaphysical system) of Leibniz's
> monadology.

Throughout his life, Gödel was an opponent of the Catholic Church; he did, however, have an abstract religious consciousness and believed in the logical provability of the existence of God. Since Immanuel Kant, we have known that no proof of a personal God is possible. But building on the ideas of Charles Hartshorne, Gödel furnished a proof of the existence of a logical God.

In early 1948 Gödel decided to seek American citizenship. In his characteristically thorough way, he began a detailed study of the U.S. Constitution in preparation for the citizenship examination. On the day before the exam, Gödel called his friend the noted economist Oskar Morgenstern, saying with great excitement and consternation that he had discovered a logical flaw in the Constitution, a loophole by which the United States could be transformed into a dictatorship. Morgenstern, who along with Einstein was to serve as one of Gödel's witnesses at the examination the next day, told him that the possibility he had uncovered was extremely hypothetical and remote. He further cautioned Gödel not to bring the matter up the next day at the interview with the judge.

The following morning Einstein, Morgenstern, and Gödel drove down to the federal courthouse in the New Jersey state capital of Trenton, where the citizenship examination was to take place. As legend has it, Einstein and Morgenstern regaled Gödel with stories and jokes on the trip from Princeton to Trenton in order to take his mind off the upcoming test. At the interview itself, the judge was suitably impressed by the sterling character and public personas of Gödel's witnesses and even broke with tradition by inviting them to sit in during the exam. The judge began by saying to Gödel, "Up to now you have held German citizenship." Gödel corrected this slight affront, noting that he was Austrian. Unfazed, the judge continued, "Anyhow, it was under an evil dictatorship ... but fortunately, that's not possible in America." With the magic word *dictatorship* out of the bag, Gödel was not to be denied, crying out, "On the contrary, I know how that can happen. And I can prove it!" By all accounts it took the efforts of not only Einstein and Morgenstern but also of the judge to calm Gödel down and prevent him from launching into a detailed and lengthy discourse about his "discovery."

This story illustrates perfectly the legalistic workings of the kind of mind that had looked into the heart of Hilbert's Program and announced to the world that the master had gone astray.

Gödel and Einstein

Gödel's reclusive nature in Princeton was a result of both health problems and an innate shyness. After living for awhile on Nassau Street in the town's center, he moved to a wood-frame house at 129 Linden Lane (later changed to 145) that he eventually purchased in 1948. He lived in this house with his wife until his death in 1978. Many German and Italian immigrant workers lived in the new neighborhood. The move had been prompted by Adele's desire to live near other German-speakers. Compared to the houses of his colleagues, such as von Neumann, Gödel's was a rather modest dwelling on the "wrong side" of town.

In part, Gödel's failure ever to leave America after his immigration reflected what Dorothy Morgenstern, wife of economist Oskar Morgenstern, described as his low opinion of Austria. She observed that "he knew what everyone knew—that many people were Nazis even before Hitler. And he knew their real opinions. And I believe he didn't feel any desire or need to go back there after [leaving]."

On these grounds, Gödel refused an honorary professorship from the University of Vienna in 1966. Although his wife went back to Europe often, Gödel constantly postponed plans to visit his aging, sickly mother in Vienna. Instead, both his brother Rudolf and his mother visited him several times in Princeton.

In the 1950s Gödel's health steadily deteriorated. His self-diagnoses became increasingly strident, his attacks of paranoia more intense. His physician, Dr. Rampona, notes that "he was actually a very difficult patient. Once I was called to his house because he was spitting blood. I diagnosed bleeding ulcer, but he refused to go to the hospital, and it was only Einstein's art of persuasion that changed his mind."

Lili Kahler, a long-term friend of the family, adds to the unsettling picture:

> He always cooked for himself, and not even Adele, who was a very good cook, was allowed to do so for him. Aside

> from the fact that he wanted to maintain his own diet, he
> was paranoid, and believed that people wanted to poison
> him. And I can only add that he almost carried things to
> their extreme.

Presumably, what she meant by "almost carried things to their extreme" is that Gödel nearly starved himself rather than eat what he thought might be tainted food.

The darkening of Gödel's spirit appears based in part on a kind of guilt that he experienced because he did not fulfill his work at the Institute. This is reflected in the following remark he once made to Dorothy Morgenstern: "No, I'm not completing the work I was meant to do at the Institute. In my role as professor, it's expected that I work more and concern myself more with other members."

The anxieties, feelings of guilt, and attacks of paranoia finally brought him to a psychiatrist, the famed Richard Huelsenbeck (known as Hulbeck in New York), previously a member of the Zurich Dada movement and friend of the painters Hans Richter and Marcel Duchamp.

It is obvious that these continuing crises led to considerable tension with Adele. She often scolded Gödel the entire day, while he remained in reclusion. She had always felt that the Institute was in the end an "old age home," and she was never really comfortable in Princeton. Hassler Whitney remembers the final period, when Adele also was rather sick:

> She would for instance say "I've never met anyone from
> the Institute and no one has phoned me." And Gödel
> would answer, "try to remember for instance this person
> with whom you chatted, that Institute member." And she
> would say, "oh no, I've never met anyone." And he would
> listen to her words and find them inaccurate, then trying
> to explain to her that they were so. But she wanted to
> express her feelings in this regard, while he could not

hear these feelings. And neither one understood where
the problem lay.

Toward the end of his life, Gödel became increasingly preoccupied with occultism. Ever since his youth,
his life and intellectual style had been permeated with a
longing for a purely spiritual, virtually immaterial existence. His library included books by Arthur Koestler
on Indian philosophy and yoga. In one such book,
Samadi is described as the final goal of yoga. Physiologi-
cally speaking, Samadi means a reduction in heartbeat,
breathing, and digestion. Spiritually, it means a state of
pure consciousness—that is, a consciousness without
volition or any content other than consciousness itself.
There is also a final Samadi to which one can voluntar-
ily ascend; it brings death of both the body and the ego
to which it is bound.

If we desire, we can give Gödel's occultism and his
Platonism a typical Viennese interpretation that is psychoanalytic in nature. In this view, Gödel was fixated on
his mother, as articulated through a denial of reality—a
radical withdrawal to the abstract world of mathematical ideas.

Gödel died on January 14, 1978, in the Princeton Hospital from "malnourishment and inanition."
Dr. Ramona, his physician, states, "He had refused all
food. He had never eaten very much, but his final
weight was only around 60 pounds. He died in the fetal
position."

However tempting the psychoanalytic interpretation may seem here, doctors note the frequency of
digestive problems in older people. The fetal position
might then be interpreted simply as a result of stomach
cramping from severe cold, itself the consequence of a
low metabolic rate insufficient to heat the body. In any

case, let us now put aside these questions and any psychological interpretations of Gödel's strange life and death. Let us consider instead the range of his intellectual achievement and the implications of his work across the spectrum of science, logic, and philosophy.

Chapter Six

Mechanism and Mathematics

Consider the following two sets of arguments:

Argument A
Everybody loves a lover.
George doesn't love himself.
Therefore George doesn't love Martha.

Argument B
Either everyone is a lover or some people are not lovers.
If everyone is a lover, Waldo certainly is a lover.
If everyone isn't, then there is at least one nonlover; call her Myrtle.
Therefore if Myrtle is a lover, everyone is.

Given the well-chronicled human proclivity for fuzzy thinking and logical inconsistency, it probably comes as a bit of a surprise for most of us to discover that both of these chains of reasoning are logically correct. Wouldn't it be nice if we had a logic machine into which we could feed these kinds of statements, turn the handle, and have the machine tell us conclusively whether the line of argument is logically valid? Such a machine is what logicians call a *decision procedure*. And it was the search for just such a procedure that sparked British mathematician Alan Turing's investigation into the foundations of computing in the 1930s. To ease our way into this story, let's focus on the ideas underlying the difference between something's being logically correct—like arguments A and B—and something's being true—like a real-world George who really does not love some real-world Martha.

In logic and mathematics, the road to truth is paved by the stones that nowadays we call a formal system. We saw the essentials of such an object in Chapter 2. Basically, a formal system consists of an alphabet of abstract symbols; a grammar for how to put sequences of such symbols together into grammatically "correct" strings; a set of axioms, which are simply strings accepted as being grammatically correct without being derived from other strings; and rules of logical inference, telling us how to create new, grammatically correct strings from existing ones.

Now suppose we are given a well-formed statement *A* and asked whether it is a logically correct consequence of the axioms of the system. We say that *A* is *provable* in the system if there is a finite chain of statements $M_1 \rightarrow M_2 \rightarrow M_3 \rightarrow \cdots \rightarrow M_n = A$, where each M_i either is one of the axioms of the system or is obtained from the previous M's via one of the rules of inference. Well-formed statements for which such a *proof sequence* exists are called the *theorems* of the formal system. We saw examples of this earlier. But just to fix the idea, here's another one: the game of chess.

A moment's thought shows that the game of chess is really a relationship between one set of abstract symbols, the Black and White pieces, and another set of abstract symbols, the squares of the board. In short, there is nothing crucial about the material embodiment of these symbols insofar as the essentials of the game are concerned. Thus, for instance, we could assign any set of symbols to represent the various playing pieces, along with another set of symbols (such as the set of positive integers 1, 2, ... , 64) to represent the squares of the board. The grammar of such a system would then specify what strings of symbols (statements) are well formed (that is, represent valid configurations of pieces on the board; for example, a black bishop cannot sit on a white square). Such grammatically correct sentences represent the possible states of play at any stage of the game. Moreover, the rules of inference of this formal system are simply the different ways in which one well-formed string of symbols can be transformed into another. In other words, the rules of inference represent the allowable moves at any stage of play. Finally, the sole axiom of the game of chess is the symbol string corresponding to the way the pieces sit on the board at the beginning of play.

Thus we see that the real world of chess pieces and playing boards can be translated into a formal-world version of the game involving only abstract symbols, rules of inference, and axioms. And the same line of argument applies to every other real-world situation that can be described in a finite number of words. We'll return to this point with a vengeance later on.

Let us again emphasize that there are two entirely different worlds being mixed up here: the purely syntactic world of the formal system and the meaningful world of mathematical objects and their properties. In each of these worlds there is a notion of truth: theorems in the formal system, factually correct statements such as "$2 + 5 = 7$" and "the sum of the angles of a triangle in a plane equals 180 degrees" in the realm of mathematical reality. As we noted in Chapter 1, the connection between the two worlds lies in the interpretation of the elements of the formal system in terms of the objects and operations of the mathematical structure. Once this dictionary has been written and the associated interpretation established, we can hope that there will be a perfect, one-to-one correspondence between the true facts of the mathematical structure and the theorems of the formal system.

We would like to deal with formal systems for which every truth translates into a theorem, and conversely. Such a system is termed *complete*. We'll consider shortly the degree to which this ideal relationship between the world of symbols and the world of mathematical facts can be approached. For the moment, however, let's stay within the formal world of symbols and rules, looking just a bit deeper into the ins and outs of computing machines and formal rule-based systems.

The problem that stimulated Alan Turing to devise the theoretical gadget now called a *Turing machine* was motivated by the Decision Problem. It can be stated as follows: For every formal system F, is it possible to find a finitely describable formal system F that "decides" any grammatically correct string in F? Loosely speaking, we ask whether there is a systematic procedure that will tell us whether any given well-formed string of the formal system F is or is not a theorem.

Turing's attack on the Decision Problem led him to construct the key element in the modern theory of computation, the Turing machine. It turns out that Turing's solution to the Decision Problem, which he carried out within the context of the theory of computation, is abstractly identical to Gödel's result in mathematical logic. So let's now turn our attention to Turing's innovative solution to the Decision Problem.

Magic Machines and Busy Beavers

What is a computation? Oddly enough, despite the fact that humans have been calculating things for thousands of years, a proper scientific answer to this seemingly straightforward query was not forthcoming until 1935. In that year Alan Turing was a student at Cambridge University, sitting in on a course of lectures in mathematical logic. A central theme of the course was the issue of whether there could exist a finite set of rules—in effect a *mechanism*—that would settle the truth or falsity of every possible statement about numbers that could be made in, say, the language of Russell and Whitehead's (in)famous work, *Principia Mathematica*. In short, the question was whether there was a machine into which we could feed any statement about numbers so that after a finite amount of time, the machine would spit out the verdict on the statement, TRUE or FALSE.

Turing's speculations about what it would mean to have such a mechanical procedure, or *effective process*, for solving this famous Decision Problem led him to develop a mathematical type of computer. This abstract gadget, the Turing machine, provided the first completely satisfactory answer to what it means to carry out a computation.

Turing took the common-sense tack of looking at what a human being actually does when carrying out a computation. It turns out that, distilled to its essence, computing comes down to the rote following of a set of rules. For example, if we want to calculate the square root of 2, we can employ the following rule for creating a set of numbers $\{x_i\}$ that will (we hope) converge to the quantity $\sqrt{2}$: $x_{n+1} = (x_n/2) + (1/x_n)$. Starting with the

initial approximation (guess) $x_0 = 1$, this rule generates the successively better approximations $x_1 = 3/2 =$
1.5, $x_2 = 17/12 = 1.4166$, $x_3 = 577/408 = 1.4142$.
After just three steps, we have the desired answer correct to four significant figures. For our purposes here,
what's important about this so-called Newton–Raphson
method for calculating the square root of 2 is not the
rapid rate of convergence but that the procedure represents a purely mechanical, step-by-step process (tech-
nically, an *algorithm*) for finding the desired quantity.

The fact that every step in such a procedure is completely and explicitly specified led Turing to believe
that it would be possible to construct a machine to
carry out the computations. Once the algorithm and
the starting point are given to the machine, computation of the sequence of results becomes a purely mechanical matter involving no judgment calls or interven-
tions by humans along the way. But it would require
a special type of machine to accomplish this computational task; not just any mechanical device would do.
A large part of Turing's genius was to show that the
very primitive type of abstract computing machine he
invented is actually the most general type of computer
imaginable. In fact, every real-life computer that's ever
been built is just a special case that materially embodies
the machine Turing dreamed up. This result is so central to understanding the limitations of machines that
it will be worth our while to take a few pages to describe
it in more detail.

The Turing Machine

A Turing machine consists of two components: (1) an
infinitely long tape ruled off into squares each of which
contain one of a finite set of symbols, and (2) a scanning head that can be in one of a finite number of

states or configurations at each step of the computational process. The head can read the squares on the tape and write one of the symbols onto each square. The behavior of the Turing machine is controlled by an algorithm, or what we now call a *program*. The program is composed of a finite number of instructions, each of which is selected from the following set of possibilities: change or retain the current state of the head; print a new symbol or keep the old symbol on the current square; move left or right one square; stop. That's it. Just seven simple possibilities. The overall situation is depicted in Figure 6.1 for a Turing machine that has 12 internal states labeled A through L. Which of the seven possible actions the head takes at any step of the process is determined by the current state of the head and by what it reads on the square it's currently scanning. But rather than continuing to speak in these abstract terms, it's simpler to run through an example in order to get the hang of how such a device operates.

Assume we have a Turing machine with three internal states, A, B, and C, and that the symbols that can be written on the tape are just the two integers 0 and 1. Now suppose we want to use this machine to carry out the addition of two whole numbers. For definiteness, we'll represent the integer n by a string of n consecutive 1s on the tape. The program shown in Table 6.1 serves

Table 6.1. A Turing machine program for addition.

	Symbol Read	
State	1	0
A	1, R, A	1, R, B
B	1, R, B	0, L, C
C	0, STOP	STOP

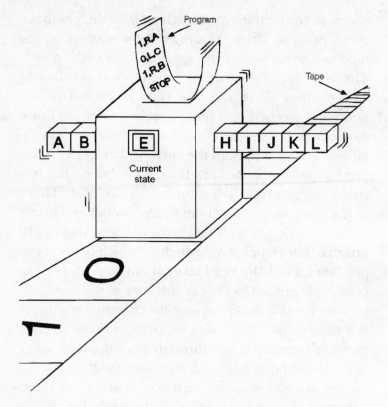

Figure 6.1. A 12-state Turing machine.

to add any two whole numbers using this 3-state Turing machine.

The reader should interpret the table entries in the following way: The first entry is the symbol the head should print, the second element is the direction the head should move, R(ight) or L(eft), and the final element is the state the head should then move into. Note that the machine stops as soon as the head goes into state C. Let's see how it works for the specific case of adding the numbers 2 and 5.

Because our interest is in using the machine to add 2 and 5, we place two 1s and five 1s on an otherwise blank (all 0s) input tape, separating them by a 0 to indicate that they are two distinct numbers. Thus the machine begins by reading the input tape

···	0	1	1	0	1	1	1	1	1	0	0	···

By convention, we assume the head starts in state A and reads the first nonzero symbol on the left. This symbol is a 1, so the program tells the machine to print a 1 on the square and move to the right, retaining its internal state A. The head is still in state A, and the current symbol read is again a 1, so the machine repeats the previous step and moves one square farther to the right. Now, for a change, the head reads a 0. The program tells the machine to print a 1, move to the right, and switch to state B. I'll leave it to the reader to complete the remaining steps of the program, verifying that when the machine finally halts, the tape ends up looking just like the foregoing input tape, except that the 0 separating 2 and 5 has been eliminated (that is, the tape will have seven 1s in a row, as required).

Before we look at the revolutionary implications of Turing's idea, we want to emphasize that Turing machines are definitely not machines in the everyday sense of being material devices. Rather, they are "paper computers" completely specified by their programs. Thus, when we use the term *machine* in what follows, the reader should read *program* or *algorithm* (that is, software) and put all notions of hardware out of sight and out of mind. This abuse of the term *machine* is implicit in Turing's idea of an *infinite* storage tape, but it's important to make the distinction as clear as possible: Turing machine = program. Period.

Modern computing devices, even home computers like the one used to write this book, look vastly

more complicated and vastly greater in their computational power than a Turing machine, with its handful of internal states and very circumscribed repertoire
of scanning-head actions. Nevertheless, this is not the
case, and a large measure of Turing's genius was that he
recognized that *any* algorithm (that is, any program)
executable on *any* computing machine—idealized or
otherwise—can also be carried out on a particular version of his machine, termed a *universal Turing machine*
(or UTM for short). Thus except for the speed of the
computation, which definitely *is* hardware-dependent,
there's no computation that a home machine can do
that can't be done with a UTM.

To specify his UTM, Turing realized that not only
the input data of the problem but also the program
itself could be coded by a series of 0s and 1s. Consequently, we can regard the program as another kind of
input data, writing it onto the tape along with the data
it is to process. Table 6.2 shows one of the many ways
in which this coding can be done.

With this key insight at hand, Turing constructed
a program that could simulate the action of any other
program P when given P as part of its input (in other
words, he created a UTM). The operation of a UTM is
simplicity itself.

Suppose we have a particular Turing machine specified by the program P. Because a Turing machine is
completely determined by its program, all we need do
is feed the program P into the UTM along with the
input data. Thereafter the UTM will simulate the action
of P on the data; there will be no recognizable difference between running the program P on the original
machine and having the UTM pretend it *is* the Turing
machine P.

Table 6.2. A coding scheme for the Turing machine language.

Program statement	Code
PRINT 0	000
PRINT 1	001
GO RIGHT	010
GO LEFT	011
GO TO STEP i IF THE CURRENT SQUARE CONTAINS 0	$1010\underbrace{0\ldots\ldots0}_{i \text{ repetitions}}1$
GO TO STEP i IF THE CURRENT SQUARE CONTAINS 1	$1101\underbrace{1\ldots\ldots1}_{i \text{ repetitions}}0$
STOP	100

What's important about the Turing machine from a theoretical point of view is that it represents a formal mathematical object. Thus, with the invention of the Turing machine, we had for the first time a well-defined notion of what it means to compute something. But this then raises a question: What exactly can we compute? In particular, is there a suitable Turing machine that will compute every number? Or do there exist numbers that are forever beyond the bounds of computation? Turing himself addressed this problem of computability in his trail-blazing 1936 paper, in which he introduced the Turing machine as a way of answering these fundamental questions.

First of all, let's be clear on what we mean by a number being computable. Put simply, an integer n is said to be *computable* if there is a Turing machine that, starting with a tape containing all 0s, will stop after a finite number of steps with the tape then containing a string of n 1s and all the rest 0s. The case of computing a real number is a bit trickier because most real numbers consist of an infinite number of digits, so we call a real number computable if there is a Turing machine that will successively print out the digits of the number, one after the other. Of course, in this case the machine will generally run on forever. With these definitions in hand, let's look at the limitations on our ability to compute numbers.

It's an easy exercise to show that for a Turing machine with n possible states of the reading head and two possible symbols that can be written on the tape, exactly $(4n+4)^{2n}$ distinct programs can be written. This means that an n-state machine can compute at most this many numbers. Letting n take on the values $n = 1, 2, 3, \ldots$, we conclude that Turing machines can calculate at most a *countable* set of numbers—that is, a set whose elements can be put into a one-to-one correspondence with a subset of the positive integers (the "counting" numbers). But there are uncountably many real numbers; hence, we come to the perhaps surprising result that the vast majority of real numbers are not computable.

This counting argument is one way, albeit a somewhat indirect one, to show the existence of uncomputable numbers. Turing himself used a more direct procedure based on what is known as *Cantor's diagonal argument*. It goes like this. Consider the following listing of names: Smith, Otway, Arquette, Bethel, Bellman, and Imhoff. Now take the first letter of the first name and advance it alphabetically by one position. This gives a T. Then do the same for the second letter of the second name, the third letter of the third name, and so on. The result is "Turing." It's clear, I think, that the name Turing could not have been on the original list, because it must differ from each entry on that list by at least one letter.

Turing's argument for the existence of uncomputable numbers follows the same line of reasoning. Suppose we list all computable numbers, written out by their decimal expansions (even though such a list will be infinitely long). Now we advance the first digit of the first number, the second digit of the second number, and, in general, the kth digit of the kth number. In this way we create a new number. This number cannot have been on the original list, because it differs in at least one position from every number on that list, but by definition, the list contains all computable numbers. Hence the new number must be uncomputable.

From the foregoing arguments, we see that uncomputable numbers are not *rara avis* in the arithmetic aviary. Quite the contrary, in fact: It's the computable numbers that are the exception rather than the rule. This surprising fact shows that all the numbers we deal with in our everyday personal and professional lives, which by their very nature must be computable, form a microscopically small subset of the set of all possible numbers. The overwhelming majority of numbers lie in a realm that's impossible to reach by following the rules of any type of computing machine. Now let's look at an amusing example of a specific uncomputable quantity.

The Busy Beaver Game

Suppose you're given an input tape filled entirely with 0s. The challenge is to write a program for an n-state Turing machine such that (1) the program eventually halts, and (2) the program prints as many 1s as possible on the tape before it stops. Obviously, the number of 1s that can be printed is a function only of n, the number of states available to the machine. Equally clear is the fact that if $n = 1$, the maximum number of 1s that can be printed is only one, a result that follows immediately from the requirement that the program cannot run on forever. If $n = 2$, it can be shown that the maximum number of 1s that can be printed before the machine halts is four. Programs that print a maximal number of 1s before halting are called *n-state Busy Beavers*. Table 6.3 gives the program for a 3-state Busy Beaver, and Figure 6.2 shows how this program can print six 1s on the tape before stopping. (*Note:* The position of the tape-scanning head is shown in boldface in the figure.)

Table 6.3. A 3-state Busy Beaver.

	Symbol Read	
State	0	1
A	1, R, B	1, L, C
B	1, L, A	1, R, B
C	1, L, B	1, STOP

Now for our uncomputable quantity. Define $BB(n)$ to be the number of 1s written by an *n*-state Busy Beaver program. Thus the Busy Beaver function $BB(n)$ is the greatest number of 1s that any halting program can write on the tape of an *n*-state Turing machine. We have already seen that $BB(1) = 1$, $BB(2) = 4$, and $BB(3) = 6$. From these results for small values of n, you might think that the function $BB(n)$ doesn't have any particularly interesting properties as n gets larger. But just as you can't judge a book by its cover (or title), you also can't judge a function from its behavior for just a few values of its argument. In fact, detailed investigation has shown that

$$BB(12) \geq 6 \times 4096^{4096^{4096^{\cdots 4096^4}}}$$

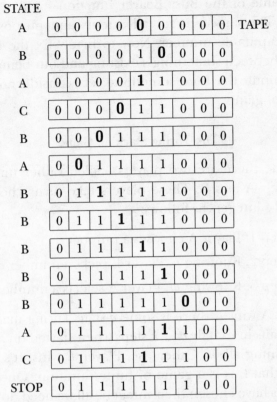

Figure 6.2. The action of a 3-state Busy Beaver.

where the number 4096 appears 166 times in the dot-ted region! Thus, in trying to calculate the value of the Busy Beaver function for a 12-state Turing machine, we quickly arrive at a number so huge that it's effectively infinite. It turns out that for large enough values of n, the value $BB(n)$ exceeds the value of *any* computable function evaluated at that same number n. In other words, the Busy Beaver function $BB(n)$ is uncomputable. Accordingly, for a concrete example of an effectively uncomputable number, just take a Turing machine with a large number of states n. Then ask for the value of the Busy Beaver function for that value of n. The answer is for all intents and purposes an uncomputable number. Now to firmly fix the difference between something's existing and our being able to compute that same something, let's consider another kind of game.

The Turing Machine Game

Assume there are two players, called rather unimaginatively A and B. These players take turns choosing positive integers as follows:

Step 1: Player A chooses a number n.

Step 2: Knowing n, Player B picks a number m.

Step 3: Knowing m, Player A selects a number k.

Player A wins if there is some n-state Turing machine that halts in exactly $m + k$ steps when started on a tape containing all 0s. Otherwise, Player B wins. It's fairly clear that this is a game of finite duration: Once the players have chosen their integers, all we need do is list the $(4n + 4)^{2n}$ Turing machines that have n states, and run each of them for exactly $m + k$ steps to determine a winner.

It's a well-known fact from game theory that any game of fixed, finite duration is determined, in the sense that there is a winning strategy for one of the players. In this case, it's Player B. Nevertheless, the Turing Machine Game is nontrivial to play, because neither player has an *algorithm* (a computable strategy) for winning the game. The proof of this fact relies on showing that *any* winning strategy involves computing a function whose values grow faster than those of the Busy Beaver function. But we already know that $BB(n)$ is uncomputable; hence, this new function must also be uncomputable. The reader should consult the References section for further details of this proof.

The Halting Problem

From our definition of computability, together with the Busy Beaver example above, it's clear that you haven't really computed anything until the computational process terminates—even for real numbers, where any finite computation generally yields only an approximation to the number you're trying to compute. This simple observation leads to a key question in the theory of computation: Is there a general procedure (an algorithm) that will tell us *in advance* whether a particular program will halt after a finite number of steps? In other words, given any Turing machine program P and a set of input data I, is there a single program that accepts P and I and will tell us whether P will halt after a finite number of steps when processing the data I? Note carefully that what we're asking for here is a *single* program that will work in *all* cases. This is the famous *Halting Problem*.

To see that the question is far from trivial, suppose we have a program P that reads a Turing machine tape and stops when it comes to the first 1. Thus in essence the program says, "Keep reading until you come to a 1, then stop." In this case, the input data I consisting entirely of 1s would result in the program stopping after the first step. On the other hand, if the input data were all 0s, then the program would never stop. Of course, in this situation we have a clear-cut procedure for deciding whether the program will halt when processing some input tape: The program will stop if the input tape contains even a single 1; otherwise, the program will run on forever. Here, then, we see an example of a halting rule that works for any data set processed by this especially primitive program.

Unfortunately, most real computer programs are vastly more complicated than this, and it's far from clear by simple inspection of the program what kinds of quantities will be computed as the program goes about its business. After all, if we knew what the program was going to compute at each step, we wouldn't have to run the program. Moreover, the stopping rule for real programs is nearly always an implicit rule of the foregoing sort, saying something like "If such and such a quantity satisfying this or that condition appears, stop; otherwise, keep computing." The essence of the Halting Problem is to ask whether there exists any *effective procedure* that we can apply to the program and its input data to tell beforehand whether the program's stopping condition will ever be satisfied. In 1936, Turing settled the matter once and for all in the negative: Given a program P and an input data set I, there is no way in general to say whether P will ever finish processing the input I.

The notion of a Turing machine finally put the idea of a computation on a solid mathematical footing, enabling us to pass from the vague, intuitive idea of an effective process to the precise, mathematically well-defined notion of an algorithm. In fact, Turing's work, along with that of the American logician Alonzo Church, forms the basis for what has come to be called the Turing–Church Thesis.

The Turing–Church Thesis

Every effective process is implementable by running a suitable program on a UTM.

The key message of the Turing-Church Thesis is the assertion that any quantity that can be computed can be computed by a suitable Turing machine. This claim is called a thesis and not a theorem because it's not really susceptible to proof. Rather, it's more in the nature of a definition, or a proposal, suggesting that we agree to equate our informal idea of carrying out a computation with the formal mathematical idea of a Turing machine.

To bring this point home more forcefully, it's helpful to draw an analogy between a Turing machine and a typewriter. A typewriter is also a primitive device, enabling us to print sequences of symbols on a piece of paper that is potentially infinite in extent. A typewriter also has only a finite number of states that it can be in: upper-case and lower-case letters, red or black ribbon, different symbol balls, and so on. Yet despite these limitations, any typewriter can be used to type *The Canterbury Tales, Alice in Wonderland,* or any other string of symbols. Of course, it might take a Chaucer or a Lewis Carroll to tell the machine what to do. But it can be done. By way of analogy, it might take a very skilled programmer to tell the Turing machine how to solve difficult computational problems. But, says the Turing–Church Thesis, the basic model—the Turing machine—suffices for every type of problem that is at all solvable by carrying out a computation.

It has probably not escaped the reader's attention that there is a striking parallel between the actions a Turing machine takes as it goes about performing a computation and the steps one follows in creating a deductive argument leading from premises to conclusions in a logical, step-by-step fashion. Turing showed the equivalence between a formal logical system and a Turing machine. In short, given any digital computer

C with unlimited memory, we can find a formal system F such that the possible outputs of C coincide with the possible theorems of F, and conversely. Using this equivalence, Turing restated the Decision Problem in computer-theoretic terms as the Halting Problem considered earlier. And because the two problems are logically equivalent, the fact that the Halting Problem has no solution implies that the Decision Problem is also unsolvable. Again, then, we run up against a brick wall in trying to get to the heart of things by following a set of rules. To hammer home this point more forcefully, here are statements of Turing's Halting Theorem and Gödel's Incompleteness Theorem, which show clearly the equivalence of the two results.

> **Gödel's Theorem.** *For any consistent formal system F purporting to settle—that is, prove or disprove— all statements of arithmetic, there exists an arithmetical proposition that can be neither proved nor disproved in this system. Therefore, the formal system F is incomplete.*

> **The Halting Theorem.** *For any Turing machine program H purporting to settle the halting or non-halting of all Turing machine programs, there exists a program P and input data I such that the program H cannot determine whether P will halt when processing the data I.*

Despite the pioneering work of Turing, which was carried out in the latter half of the 1930s, it was really the work of Kurt Gödel a few years earlier that dealt the coup de grace to the idea that there was no meaningful difference between the real-world notion of a truth and the formal-system concept of proof. Let's now turn to a consideration of what Gödel did and what it means for the hope of understanding the world by following rules.

Truth is Stranger Than Proof

In his book *Infinity and the Mind,* mathematician and science fiction writer Rudy Rucker recounts his visit to a church in Rome outside of which stands a huge stone disc. Carved onto this disc is the face of a hairy, bearded man whose slot-shaped mouth is located about waist level. According to popular legend, God has decreed that anyone who sticks a hand into the mouth and then utters a false statement will never be able to pull the hand back out again. Rucker states that he went to this church, stuck his hand into the mouth, and said, "I will not be able to pull my hand back out again." Needless to say, Rucker left Rome with all appendages intact. This story illustrates the logical basis for why it will never be possible to produce a "universal truth machine" capable of generating all possible real-world truths.

Suppose such a universal truth machine, UTM, does indeed exist. (We use the same acronym, UTM, for this hypothetical gadget as for the universal Turing machine for reasons that will become apparent as we proceed.) Now feed the following statement S into the machine: "The UTM will never print out this statement." If the UTM ever does print out S, then S will be false, so the UTM will have printed out a false statement. But this is impossible because, by assumption, the UTM is supposed to print out only true statements. Therefore, the UTM will never print out S, which implies that S is indeed a true statement. But now we have a true statement that the UTM will never print out, contradicting the fact that the machine is universal (that is, that it will print out *all* true statements). The perceptive reader will recognize the self-referential sentence *S* as another version of the famous Epimenides Paradox discussed in Chapter 2 of this book.

The punch line of the preceding argument is that truth is not finitely describable. Perhaps it's not too surprising that there is no set of rules sufficient to generate all possible real-world truths. After all, the rules themselves exist within the world they are supposed to describe, so it's a little like pulling yourself up by your own bootstraps to ask for a finite set of rules that will generate the infinity of all possible real-world truths. What is surprising, however, is Gödel's proof that this same limitation holds for the much smaller and far less cluttered world of the whole numbers.

At first glance, this implication of Gödel's result seems to sound the death knell for the idea of a "thinking machine." After all, if there are truths that the human mind can know but that cannot be accessed by a computing machine, how can we ever hope to construct a machine "intelligence" that can duplicate human cognitive processes? But things are never as simple as they seem. The next chapter shows why.

Chapter Seven

Thinking Machines and the Logic of Incompleteness

In 1989, the theoretical physicist Roger Penrose published *The Emperor's New Mind,* a book whose central argument is that the human mind is capable of transcending rational thought, and hence can never be duplicated in a machine. Before going on, note that here we are using the term *rational thought* in the strong sense of following rules or an algorithm to arrive at a result by a *rational* process of logical inference. Thus there is no connection here with the everyday notion of rationality as related to self-interest or prudent action. This message, which Penrose justified via a wildly speculative appeal to quantum processes that we'll consider in more detail later, surely came as a great source of comfort to many computer-phobes and anti-AI types, a fact that doubtlessly explains the appearance of such a technical book on the best-seller lists for months on end. Be that as it may, Penrose's anti-AI argument is that mind is somehow bigger than rational thought.

A key ingredient in Penrose's argument is Gödel's famous result showing that there are true statements of arithmetic that the human mind can know but that cannot be the end result of following a fixed set of rules—that is, a computer program. Although there are well-known reasons why Gödel's Theorem should be taken with several shakers full of salt when it is used it as an argument against thinking machines, what's important about Gödel's result for our purposes here is that it suggests that there are indeed limits to the rational powers of the human mind. The big question for mechanists then becomes "Can these limits be removed, or at least extended?" As a prologue to confronting this issue head-on, let's first outline briefly the basic structure of the field of cognitive science, of which artificial intelligence is a part. We will then shift to a more detailed investigation of AI itself and to the role that Gödel's result plays in the philosophical debate over thinking machines.

The Dancer and the Dance

An old joke that professors of philosophy use to wake up their students goes like this: "What is mind? No matter. What is matter? Never mind." It's pretty feeble as jokes go, even by the rather low standards of academia, but this epigram illustrates the very important point that although it's pretty much agreed that a material object—the human brain—gives rise somehow to the mind, the mind itself seems to have no material composition. Rather, it appears to consist solely of patterns of information existing in some realm beyond ordinary space and time. But with such a picture in mind (no pun intended), how can we even begin to address the foundational question of the nature of knowledge and how it is represented in the human mind? This is the central issue underlying what has in recent years come to be termed *cognitive science.*

In their wilder moments, practitioners of the cognitive sciences have been heard to murmur that their field represents the vanguard of twenty-first-century science. Just as the development of relativity theory and that of quantum mechanics were the defining events of twentieth-century science, the study of the body and that of the brain will be the focal points of science in the coming century. Or so they say. But studies of the brain, mind, and knowledge are the traditional province of such fields as psychology and philosophy. What has cognitive science added to these disciplinary approaches that merits such wild-eyed enthusiasm? To answer this question, it's necessary to take a harder look at what the term *cognitive science* means.

Harvard psychologist Howard Gardner contends that cognitive science is an amalgam of the six traditional academic disciplines of philosophy, psychology, linguistics, artificial intelligence, anthropology, and neuroscience, linked—strongly or weakly—in accordance with the chart shown in Figure 7.1. In this figure, the black lines indicate strong interdisciplinary ties, the dashed lines weak ones.

In addition to focusing on the interdisciplinary aspect of the cognitive sciences, Gardner identifies five features, or "fingerprints," that distinguish the cognitive sciences from their disciplinary origins. These characteristics represent both the core assumptions of the field and the unifying methodological approaches that separate the cognitive scientists from their more disciplinary-oriented colleagues. It's worth listing these fingerprints as a starting point for our subsequent discussion.

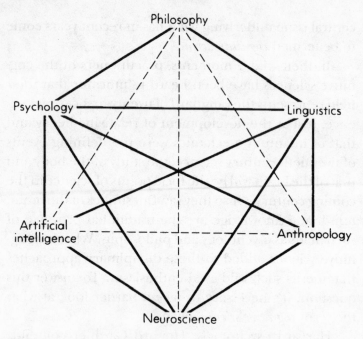

Figure 7.1. Interconnections among the cognitive sciences.

• *Representations*. Cognitive scientists believe that in speaking about human cognitive activities, it is necessary to talk about mental representations and to assume a level of analysis separate from the biological, on the one hand, and the sociocultural, on the other.

• *Computers*. Central to any understanding of the human mind is the digital computer. Not only are computers needed to carry out studies of various sorts, but, more important, the computer serves as the best available model for how the human mind actually functions. Thus, although a cognitive scientist may not actually make use of a computer in a hands-on manner, she or he believes that the computer *as an information-processing device* is the right metaphor for studying the operations of the brain and the mind.

- *De-emphasis of Affect, Context, Culture, and History.* Certain factors, such as the influence of emotions and the contribution of historical and cultural inputs, may be important for cognitive functioning. But cognitive science de-emphasizes these factors as being of secondary importance for understanding the working of the mind and, hence, regards them as unnecessary complications.

- *Interdisciplinary.* The cognitive scientist believes that there is much to be gained from interdisciplinary approaches to the mind. Therefore, even though most practitioners are drawn from one (or more) of the six disciplinary areas noted above, there is the hope that some day the boundaries between these disciplines will be diminished or perhaps will even disappear.

- *Classical Philosophical Problems.* The agenda of issues and the set of concerns of the cognitive scientist are strongly rooted in problems that have long concerned epistemologists in Western philosophy. Thus the modern cognitive scientist grapples in his or her own way with many of the same questions surrounding mind and knowledge that Plato addressed in the *Meno*, Descartes in his *Cogito*, and Kant in *The Critique of Pure Reason*.

Now that we are acquainted with these guidelines to what constitutes the field, it's almost irresistible to ask about the degree to which cognitive scientists have been able to make meaningful headway in understanding how the human mind processes information into knowledge. Is cognitive science really the vanguard of an intellectual revolution, as its adherents claim? Or is it merely (as more than one critic has argued) a second-rate scholars' refuge from the traditional areas? It may be premature to try to assess the accomplishments of a field that has been in existence only for a decade or so, but it still seems worthwhile from the standpoint of pinpointing gaps in the research agenda, as well as incipient limitations arising from results already in hand.

Clearly, to do justice to even a small fraction of the work being done under the cognitive science masthead is far beyond the scope of this chapter. Therefore, as a surrogate for the field as a whole, let us focus on the subfield of cognitive science that is most closely associated with Gödel's work: the problem of artificial intelligence. By examining the arguments for and against the possibility of building a computing machine that can think like you and me, perhaps we can sneak up on the question of whether the cognitive sciences are likely to deliver on the promise of a scientific theory of mind in the coming century. And we should be able, at the same time, to illuminate the absolutely central role that Gödel's Theorem plays in this venture.

Real Brains, Artificial Minds

In 1950 Alan Turing published the paper "Computing Machinery and Intelligence," which sparked a debate that rages to this day: Can a machine think? In addition to its seminal role in drawing attention to the question of machine intelligence, Turing's paper was notable for its introduction of an operational test for deciding whether a machine really was thinking—human-style. This criterion, now termed the "Turing test," is unabashedly behavioristic in nature, involving the machine fooling a human interrogator into thinking it is actually a human. Turing's rationale for proposing what he called the Imitation Game was that the only way we can decide whether other humans are thinking is to observe their behavior. And if this criterion is good enough to use to decide whether humans are thinking, then fairness to machines dictates that we should apply the same criterion to them.

Amusingly, on November 8, 1991, the Boston Computer Museum held the world's first hands-on Turing test, in which eight programs conversed with human inquisitors on a restricted range of topics that included women's clothing, romantic relationships, and Burgundy wine. At the day's end, the judges awarded first prize to a program called *PC Therapist III*, which was designed to engage its questioner in a whimsical conversation about nothing in particular.

For example, at one point the program suggested to a judge, "Perhaps you're not getting enough affection from your partner in the relationship."

The judge replied, "What are the key elements that are important in relationships in order to prevent conflict or problems?"

"I think you don't think I think," responded the terminal.

This kind of interchange did little to fool the judges, most of whom said they were able to spot the "commonsense" mistakes that were an immediate giveaway separating the computer programs from the humans. Nevertheless, the overall conclusion from this historic experiment is that perhaps the Turing test isn't as difficult as many people originally thought, because even the primitive programs in this contest managed to fool some of the judges.

A telling argument against the adequacy of the Turing test as a benchmark of intelligence has been advanced by philosopher Ned Block. Suppose, Block argues, that we write down a tree structure in which every possible conversation of less than five hours in duration is explicitly mapped out. This structure would clearly be enormous—much larger than any existing computer could store. But for the sake of argument, let's ignore that difficulty and put this tree structure into the computer.

By following our tree structure, the machine would interact with its interrogator in a way indistinguishable from the way in which an intelligent human being would do so. Yet the machine merely makes its way through this simple tree, which strongly suggests that the machine has no mental states at all. And this same conclusion holds for any conversation of finite duration.

From this argument, Block draws the moral that thinking is not fully captured by the Turing test. What's wrong with the tree structure is not the behaviors it produces. No, the problem is the *way* it produces them. Intelligence is not just the ability to answer questions in a way indistinguishable from the way an intelligent person would answer them. Rather, to call a behavior "intelligent" is to make a statement about *how that behavior is produced.*

Considered in terms of whether it adopts a first-person or a third-person point of view, the Turing test clearly represents a third-party perspective on human intelligence. Standing outside the system, the test is designed to discern human intelligence in a machine by observing only the behavioral output of the machine. The Turing test says nothing about the internal constitution of the machine, how its program is structured, the architecture of the processing unit, or its material composition. In Turing's view of intelligence, only behavior counts. And if you have "the right stuff," then you are a thinking machine.

Let's now turn to a short account of the arguments of both the pro-AI and the anti-AI factions. (For more details on these competing schools of thought on AI, the reader is invited to consult the accounts given in the References.)

Pro-AI

The pro-AI world is roughly divided into two basic camps, Top Down and Bottom Up. Members of the first group see the basic hardware of the brain as irrelevant to the issue of duplicating human intelligence in a computing machine. Consequently, their attempts to capture the results the brain gets center on trying to abstract the rules that the brain uses and then code these rules into a form congenial to computing machines.

Bottom Uppers, on the other hand, argue that perhaps the way our particular human type of brain is physically constituted plays a crucial role in our cognitive abilities. If so, the argument goes, then it's impossible to capture cognition—human-style—in a machine without respecting this physical structure. Thus these "New Connectionists" focus their attempts to mimic the mind in a machine by constructing programs whose functional organization mirrors as closely as possible that of the human brain. By way of prelude, we shall take a slightly longer look at the details of these two very different approaches to the problem of capturing human thought in a machine. First, the Top Downer's view of the world.

The key words in the Top Down vocabulary are *representations* and *rules*. Since the very first Top Down program (the General Problem Solver created by Herbert Simon, Alan Newell, and Cliff Shaw in the 1950s), the twin problems confronting these research efforts have revolved around how knowledge is to be represented in symbols and what rules should be used to combine these symbol strings into new, cognitively meaningful strings. This description makes it clear that there is no place, in the research manifesto of the Top Down view of AI, for the actual neurophysiological hardware of a real brain. Rather, the time and energy of Top Down researchers is devoted to seeking clever representation schemes and what might loosely be called the "rules of thought." In short, these research agendas are focused on skimming off the symbolic representation schemes and rules of thought used by the brain, utterly ignoring the brain's actual hardware. Top Down work has been divided by Hubert and Stuart Dreyfus into three rather distinct phases:

• *Representation and search (1955–1965)*. During this period, work centered on showing how a computer can solve certain classes of problems using the general heuristic search technique termed "means–end analysis." This involves making use of any available operation that reduces the distance between the current state of the system and the description of the desired goal. Simon and Newell made extensive use of these ideas, abstracting the heuristic technique for incorporation into their General Problem Solver.

• *Microworlds (1965–1975)*. Early on, the Top Down approach scored some seemingly impressive victories, especially in severely restricted arenas such as proving geometric theorems, playing chess, and other areas in which the problems could be solved via the combination of a large amount of formal logical manipulations and a minimal amount of real-world background knowledge. Unfortunately, it soon became clear that most everyday human problem solving does not involve problems that exhibit this happy conjunction of features. Experience with machine language translation revealed that most human cognition involves a considerable amount of background knowledge about the world—what many have termed "tacit knowledge." For example, one Russian–English translation program translated the English idiomatic phrase "The spirit is willing but the flesh is weak" into the Russian equivalent of "The vodka is good but the meat is rotten." With these kinds of problems emerging at an ever-increasing rate as researchers tried to create programs for practical, everyday tasks, the challenge for Top Downers became how to account for the necessary background knowledge in their rules and representations.

One early attempt was a kind of AI version of the age-old Procrustean fit—namely, create an artificial world inside the machine, a world every aspect of which the machine has complete knowledge about. Of course, these artificial worlds, or, as they came to be termed, "microworlds," are vastly slimmed-down versions of the real thing. But the hope was that after those features of the real world deemed important for a given task had been dramatically abstracted, the machine could then be given enough background information about this abstract world to be able to think intelligently about objects and relationships in these slimmed-down, artificial worlds.

Unfortunately, microworlds are not worlds but isolated, meaningless domains. And researchers gradually came to the conclusion that there was no way such domains could be combined and extended to encompass the many worlds of daily life.

● *Common-sense knowledge (1975–present).* The two previous periods of Top Down AI were characterized by efforts to determine how much could be done with as little knowledge as possible. But the problem of common-sense knowledge could not be swept under the rug forever. The obvious next step was to try to introduce data structures for stereotyped situations as a way to incorporate everyday, taken-for-granted knowledge into computer programs. But with the failure of these kinds of approaches (things like Marvin Minsky's "frames" and Roger Schank's "scripts"), it finally became clear that a radically new approach was needed. It was time to give up on the conviction of Descartes, Husserl, and the early Wittgenstein that the only way to produce intelligent behavior is to mirror the world with a formal theory in the mind. At this point, classical Top Down, symbol-based AI became an example of what Imré Lakatos has called a "degenerating research program." Enter the Bottom Up view of the world.

Looking at the brain from the other end of the telescope, Bottom Up proponents argue that the physical hardware of the brain *does* matter when it comes to human cognition. And if we're to have any hope of duplicating that kind of intelligence in a machine, it behooves us to account explicitly for the structure of that hardware in our programs. What this means is that we need to take an "insider's" perspective, looking at how the brain is actually wired up and how that structure serves to generate the kind of observable behavior we deem "intelligent."

Here are a handful of features that just about every neurophysiologist and connectionist agrees are characteristic aspects of the human brain:

- *Simple processors.* Most of the work of the brain is done by an unimaginably large number of neurons, each of which, taken by itself, is an ultraprimitive computer not much more complicated than a simple "on–off" switch.

- *Massive parallelism.* The ten billion or so neurons of the brain are connected via a network of axons and synapses, a scheme that results in concurrent operation of the neurons. Thus, if we imagined each neuron as a light bulb, a motion picture of the brain in operation would show an array of billions of lights flashing on and off in a bewildering variety of patterns. This picture would look much the same as a Times Square message board, consisting of many rows of individual flashing lights that, taken together, form a recognizable pattern. The problem is that at present, we haven't the foggiest idea what the brain's patterns mean.

● *Unprogrammed*. In contrast to modern digital computers, whose program of instructions is very inflexible and "brittle," the brain seems to be relatively unprogrammed. Rather, the strengths of the synaptic connections, which determine the firing pattern of the neurons, seem to be governed more by various learning procedures in the brain than by any kind of direct setting by a "programmer."

● *Adaptable*. The connective pattern in the brain is very plastic, allowing the brain to, in essence, reprogram itself. This, in turn, provides the basis for things like memory, learning, and creative thought. The necessity of plasticity for memory and learning is clear; both processes involve the brain changing its state in some semipermanent fashion. Moreover, because by definition a creative thought represents something new that's generated and stored in the brain, it too must arise as a result of the brain's ability to reconfigure its neuronal connections somehow. Of course, this is only a *necessary* condition for creative thought; *sufficiency* remains as great a mystery as ever.

Using these desiderata as a checklist, Bottom Up proponents of AI have been busying themselves with the development of neural networks aimed at mimicking in one way or another the hardware of the brain. The general structure of such a network is shown in Figure 7.2. The basic idea is to associate each processing element in the net with a physical neuron in the brain. Each of the connections between these elements has a numerical weight that can be either positive or negative. The size of the weight controls the influence that one element has on another, a positive connection being able to excite an element and a negative one to inhibit it. Overall, the activation of an element is then determined by a combination of the excitatory and inhibitory influences it receives from its neighbors. Such a net is next trained to respond to a set of patterns by adjusting these weights. Once the net has been trained, the network is able to respond "intelligently" to other patterns that have some features in common with the training set.

Hence, instead of dictating the rules of thought by fiat, a Bottom Up theorist sees whatever rules of thought the machine embodies as arising in an emergent way from the reconfiguration of the pattern connecting the many processors as the machine learns to survive in its environment. In this essentially internal view of cognition, the machine starts out with very little knowledge of the world but with a hardware configuration that's pretty plastic. As time goes on and the machine interacts with its environment, certain patterns of thought prove more effective than others at solving the problems of daily life. These successful patterns then ultimately get wired into the connections linking the machine's processors. It is these connections that determine whatever rules we can say the machine uses in arriving at its actions and decisions.

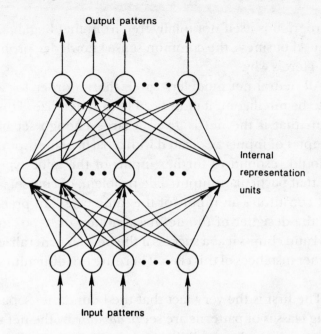

Output patterns

Internal
representation
units

Input patterns

Figure 7.2. Structure of a prototypical neural network.

A neural network trained to respond to a set of patterns displays many of the features we associate with the human brain. For example, if only a part of one of the target patterns is shown, the net is still able to recognize it. Moreover, the performance of the net degrades gradually as individual neurons "misfire" or are removed from the net. And the network can also recognize novelty in the sense that it can recognize patterns that are not part of the training set but have features in common with the target patterns. All in all, a Bottom Up neural net represents a view of the world that is theory-free, suggesting that it is possible to behave intelligently in the world without having a theory of that world. But how much of everyday intelligent behavior can such a network be expected to capture? Are there any limits on what a well-trained network is capable of? As it turns out, the Bottom Up connectionist program is itself not totally free from that bugaboo of the AI business, the common-sense knowledge problem. Here's why.

All neural net modelers agree that in order for a net to be intelligent, it must be able to generalize. This means that if the net is given a large enough set of examples of inputs associated with a particular output, it should respond to further inputs of the same type with that particular output. The problem, of course, is what constitutes an input "of the same type?" In practice, the designer of the net has a notion of "type" in mind and counts it as a success if the net can generalize to other instances of this class. There are two difficulties here.

The first is the very fact that the boundaries separating classes of patterns are set *in advance* by the net's designer. Thus the possibility for novel forms of behavior, which is clearly part of what we call human intelligence, is severely restricted. The second problem arises when the net produces an unexpected response to a given input. Can we really say that the net has failed to generalize? Perhaps the net has been acting all along on a different idea of "type," and this unexpected association has just revealed that difference.

In light of these difficulties, we might conclude that in order for a neural net to share our human sense of generalization, it must also share the size, connective structure, and initial configuration of the brain. It must also share our idea of what constitutes an appropriate output, which suggests that it must share our needs, desires, and emotions and must have a human-like body with appropriate physical abilities for movement, sensory inputs, and the like. If this indeed turns out to be the case, then the Bottom Up program will have foundered on exactly the same rock as the Top Downers, the rock of common-sense knowledge. To quote Hubert and Stuart Dreyfus,

> If the minimum unit of analysis is that of a whole organism geared into a whole cultural world, neural nets as well as symbolically programmed computers still have a very long way to go.

The one thing that both Top Down and Bottom Up devotees agree on is the claim that there is no obstacle, in principle, to the duplication of human cognitive capacity in a machine. The dispute revolves only about how to go about duplicating it. But an impressive array of arguments have been marshaled against the very idea of such "strong AI." Let us now give the floor to those who maintain that machines will never think like you and me.

Anti-AI

Whereas the majority of pro-AI advocates are computer scientists, psychologists, mathematicians, and others of that ilk, most of the arguments against AI have been put forward by philosophers. These cries of outrage against the very idea of strong AI seem to be based on one of three main lines of attack: phenomenology, anti-behaviorism, and Gödel's Theorem. Let's spend a page or two describing each of them.

● *Phenomenology*. One of the most popular spokesmen for the anti-AI cause has been Hubert Dreyfus, a philosopher at the University of California at Berkeley. Dreyfus, along with his brother Stuart, a professor of engineering who is also at Berkeley, argued against the possibility of strong AI by appealing to the works of the phenomenological philosophers Heidegger, Husserl, and Merlau-Ponty. These giants of modern continental philosophy claimed that there are many human cognitive activities that simply cannot be thought of as the following of a set of rules. A favorite example of the brothers Dreyfus in this regard involves learning how to drive an automobile.

According to the Dreyfuses, gaining expertise at driving a car involves passing through five successive identifiable stages:

Novice: At this lowest skill level, context-free rules for good driving are acquired. Thus one learns at what speed to shift gears and at what distance it's safe to follow another car at a given speed. Such rules ignore context-sensitive features such as traffic density and weather conditions.

Advanced beginner: Through practical on-the-road experience, the novice driver learns to recognize concrete situations that cannot be described by an instructor in objective, context-free terms. For instance, the advanced beginning driver learns to use engine sounds as well as the context-free speed as a guide for when to shift gears and learns to distinguish the erratic behavior of a drunk driver from the impatient actions of an aggressive driver in a hurry.

Competence: The competent driver begins to superimpose an overall driving strategy upon the general rule-following behavior of the novice and the advanced beginner. He or she no longer merely follows rules that permit safe and courteous operation of the car but drives with a goal in mind. To achieve this goal, the competent driver may now follow more closely than normal, drive faster than is allowed, or in other ways depart from the fixed rules learned earlier.

Proficiency: At the previous levels, all decisions were made on the basis of deliberative, conscious choices. But the proficient driver goes one step further and makes decisions on the basis of a "feel" for the situation. There is no deliberation; things just happen. Thus, for example, when attempting to change lanes on a busy freeway, the proficient driver may instinctively realize that there's another car coming up on the blind side and delay making a move. This instinctive reaction may arise out of experience in similar situations in the past and memories of them, although to an outside observer it may appear an unexplainable "lucky guess." Somehow there is a spontaneous understanding or "seeing" of a plan or strategy.

Expert: An expert driver no longer sees driving as a sequence of problems to solve, nor does he or she worry about the future and devise plans. Such a driver simply becomes one with her or his car and registers the experience as just driving rather than as driving a car. Thus expert drivers have an intuitive understanding of what to do in a given setting. They neither solve problems nor make decisions; they just do what normally works.

The moral of this fable in five parts is the *claim* that there is more to intelligence and expertise than mere calculative rationality. Expertise doesn't necessarily involve inference; the expert sees what to do *without* applying rules. This is the essence of the Dreyfus argument against the possibility of a rule-based program's ever achieving anything that even remotely approximates genuine human intelligence. Of course, this is simply speculation on the Dreyfuses' part that behavior is not rule-based; it is hardly a theorem.

It's pretty clear from this example that the Dreyfus position against strong AI is essentially a third-person argument against rule-based behavior. The claim is that simply by looking at the external behavior of a human being, we can see cognitive activity that cannot be attributed to the following of a set of rules. The Dreyfuses then claim that it is impossible to program a computer to drive a car in a manner indistinguishable from an expert human driver. The computer would have to have a set of rules to do this task, but there are no such high-level rules that human drivers follow in guiding their vehicles through traffic congestion and high-speed freeway traffic. Because machines can only follow the rules encoded into their programs, the full experience of even such a relatively simple human task as driving a car cannot be captured within the confines of a computer program. Ergo, machines cannot think—at least not like you and me.

As an aside, it's of some significance here to note that Hubert Dreyfus admitted to one of the authors (JC) in a private conversation that the main thrust of his anti-AI argument is directed against the Top Down approach to strong AI and that it may well be possible to duplicate human capacity in driving (and everything else) by following a Bottom Up approach. But even this approach may run aground on the shoals of the same common-sense knowledge reef, as we have already noted. Now let's consider another class of reasons why machines will never think.

• *Anti-behaviorism.* One of the strongest arguments yet mounted by the philosophers against strong AI has been articulated by John Searle, also a professor at Berkeley. Searle's argument is essentially a first-person claim that what goes on inside a computing machine when it moves symbolic representations around in accordance with a program is pure syntax. But, Searle argues, no amount of syntax alone (that is, symbol shuffling, can ever give rise to semantics. In other words, the computer can have no understanding of the *meaning* of the symbols it manipulates. And without meaning there is no intelligence.

To dramatize this first-person point of view on thinking, Searle constructed the colorful analogy of what is now called the Chinese Room argument. This involves imagining someone ignorant of Chinese, who is locked up in a closed room with a dictionary containing only Chinese ideographs, together with a set of cards each of which has a Chinese character printed on it. The person receives similar cards with Chinese characters through a slot in the door to the room. He or she then looks up the character on the card in the dictionary and then passes back out, through the slot, the card containing the character called for by what she or he finds in the dictionary listing.

It's clear that from the perspective of the man in the room, there is no understanding of Chinese here at all; that is, there is no semantics. There is only a purely syntactic shuffling of cards back and forth through the slot in accordance with rules dictated by the dictionary. But from the third-person, essentially behaviorist perspective of a native Chinese speaker outside the room, the sequence of cards flowing into and out of the slot may well be seen as a perfectly sensible written conversation in Chinese about, say, tomorrow's weather, the state of the stock market, or the end of the world. Searle's point here is that the actions of the person in the room duplicate exactly what happens inside a computer as it goes about its business of transforming input symbol strings into output strings.

When Searle first published this anti-AI argument in 1980, the howls of outrage from the pro-AI community could be heard from Stanford to MIT and back again. For Searle's response to these quibbles, as well as many more details of the arguments themselves, the reader should consult the peer commentary accompanying the original article cited in the chapter references. On this inconclusive note, let's move on to our final class of anti-AI claims, the ones that are of most interest to us in this book. These are the arguments that rest on an appeal to Gödel's famous Incompleteness Theorem.

• *Gödel's Theorem.* One of the most influential arguments against the possibility of strong AI was advanced in 1961 by Oxford philosopher John Lucas, who invoked Gödel's result, saying, in effect, that because there exist arithmetical truths that we humans can see to be true but that a machine cannot prove, the capacity of the human mind must transcend that of any machine. As noted at the beginning of the chapter, Roger Penrose, another Oxford don, has recently appealed to much the same line of argument to conclude that machines cannot think as humans do. Penrose adds a twist to the usual party line by speculating that at least some part of human thought involves making contact with uncomputable quantities. And the best way he can come up with to explain how this might happen is to posit mysterious quantum events influencing the brain's neuronal firing patterns.

For over thirty years now, the arguments have raged hot and heavy against the Gödelian line of reasoning against strong AI, and we don't want to bore the reader by going into them again here. They can all be found in many places, including some of the volumes cited in the References. Suffice it to say for now that we have already seen that Gödel's theorem involves certain assumptions. The most important of these is that the formal system (the computer program) be logically consistent. It's rather clear that satisfaction of this condition by the human mind is a dubious proposition, at best; we can all remember instances when we behaved in a demonstrably inconsistent manner. And if the system is logically inconsistent, all bets are off as far as appeals to Gödel's result are concerned.

With all these arguments casting doubt on the feasibility of both the Top Down and the Bottom Up programs for strong AI, is it really plausible that we'll ever directly be able to *design* a "brain"? After all, Gödel's result tells us that there are limits to what we can do by way of rationally planning anything. Perhaps both the symbol-processing and the connectionist exercises are doomed to failure from the beginning. Maybe the brain is just too complex for us ever to fully understand, let alone design a machine to duplicate. Maybe. But even if this does indeed turn out to be the case, all is not lost. Read on.

Minds, Machines, and Evolution

On January 3, 1990, Tom Ray, a naturalist from the University of Delaware, pushed the start button on his computer, thereby kicking into action a program called Tierra. Letting the program run all night, in the morning Ray found an electronic ecosystem of dazzling diversity—populations of many different types of organisms in his machine, all of which descended from a single ancestral organism that Ray had inserted into the program to get it started. As Ray put it, "From the most basic instructions there emerged an astonishing complexity." Such are the powers of evolution.

The Tierra simulator is an attempt to mimic the process of Darwinian evolution in a machine. The

organisms in this electronic ecosystem are self-repro-
ducing strings of IBM assembler-language code. Each
of these programs competes against the others for memory locations in the machine, so there is no *a priori* criterion imposed from the outside as to what is a "fit" organism. What's fit or unfit changes over time, depending
on how the organisms in the "soup" mutate, recom-
bine, and, in general, evolve so as to leave as many
copies of themselves in the machine as possible. For
details of the many clever ways Ray developed to ensure
that a population of his "critters" does indeed capture
the characteristic features of a real ecosystem, the reader is referred to Ray's account of the whole experiment
given in his papers cited in the References.

The Tierra exercise was the first ever to definitively
demonstrate that the process of evolution is independent of a particular material substrate. It can take place
just as easily among a population of computer programs
competing for memory space in a machine as among
populations of carbon-based organisms competing to
survive in an earthly environment. So why couldn't it
happen in a population of machines?

Interestingly enough, this evolutionary argument
seems to be the position favored by Gödel himself.
When asked whether his theorem was an insurmount-
able barrier to the development of a true mechanical
intelligence, Gödel's response was

> ... it remains possible that there may exist (and even
> be empirically discoverable) a theorem-proving machine
> which in fact *is* equivalent in mathematical intuition [to
> the human mind], but cannot be *proved* to be so, nor
> even be proved to yield only *correct* theorems of finitary
> number theory.

By this remark, Gödel is suggesting that a machine
equivalent in brainpower to the human mind might

actually be created—for example, by evolution—but that if such a device did exist, we would never understand it. It would be too complex for us.

Gödel's prescription, then, is not to build a brain but rather to *grow* one! And Ray's experiment shows that there is no logical barrier to following this dictum. Thus what both Gödel and Ray are saying, in effect, is that a machine equal to humans in cognitive capacity will be just an example, though a very special one, of what we can call artificial life (AL). Because this whole theme has become a hot topic of late, let's conclude this account of AI and Gödel by exploring just a few of the striking parallels between the research agendas of the AI'ers and the AL'ers by examining the following set of hypotheses put forth by Steen Rasmussen, which underlie belief in the very existence of AL.

Postulate 1: *A universal Turing machine can simulate any physical process.* This is the assumption that the information transmission rules of any physical process can be mimicked by a suitably programmed computer. In short, the Turing–Church Thesis is true for physical systems.

Comment: Interestingly enough, this is exactly the assumption that Roger Penrose has called into question in his treatment of strong AI. An important part of Penrose's anti-AI argument is that the brain has ways of processing information that transcend computability as characterized by a universal Turing machine. On the other hand, weakening this postulate to the statement that all human cognitive activity is computable is exactly the assumption sustaining the belief of pro-AI researchers in the ultimate success of their strong-AI research programs.

Postulate 2: *Life is a physical process.* The crucial point here is the claim that life is a consequence of the functional organization of the different parts of a system and that these functional aspects can be created in many different types of physical hardware. In particular, the relevant functional properties that give rise to life can be created in a computing machine.

Comment: Substituting the word *cognition* for the word *life* in the above statement leads to the functionalist position on strong AI. Thus, whether you're a Top Down or a Bottom Up AI man or woman, there is nothing about thinking that transcends ordinary neurophysiological processes in the brain. Human thought is then a consequence of how the physical components of the brain are organized and does not in any way involve the exact details of the hardware. Of course, connectionists hold that the wiring pattern linking the brain's neurons is important, but they believe that pattern can be duplicated in many distinct sorts of actual hardware—including a digital computer.

Postulate 3: *There are criteria by which we can distinguish between living and nonliving systems.* Even though all presently known conditions for life seem rather fuzzy, this postulate asserts that agreement can in principle be reached as to what is and what isn't alive. In particular, all living systems should include the functional activities of metabolism, self-repair, and replication.

Comment: In the AI context, this postulate gives rise to things like the Turing test and Chinese Room arguments. How do we know someone is thinking? How do we know something is alive? In both cases there seem to be pretty strongly held intuitive ideas enabling us to say, in effect, "I know it when I see it." But the fur starts to fly when it comes to giving an explicit set of criteria that apply in all cases.

Postulate 4: *An artificial organism must perceive a reality R^*, which for it is just as real as the "real" reality R is for us.* An important consequence of this assumption is

Postulate 5: *The realities R^* and R have the same ontological status.* In other words, what we call reality is no more or less real than the reality seen by an artificial organism in a machine.

Comment: Acceptance of these two assumptions in either the AI or the AL world leads immediately to a plethora of issues surrounding "rights" for machines. If a genuine thinking machine has the same ontological status of a thinking human, then it's hard to make a case against giving such a device the same civil rights we accord to humans. Or so goes the argument, anyway.

Postulate 6: *We can learn about the fundamental properties of our reality R by studying the details of different R^*s.* This means that looking at what artificial life does inside a machine can give us insight into what our human form of life is doing outside the computer.

Comment: This postulate is the *raison d'être* for the entire cognitive science undertaking. If we accept the machine version of neurons, thoughts, language, or whatever as a valid representation of that same concept in R, it follows that the machine-world version and the real-world version are "isomorphic." In other words, they are functionally equivalent, and whatever you learn from the study of one can be transferred to the other, necessary changes having been made. For example, this line of reasoning is what supports most of the interest in Tom Ray's Tierra simulator as a way of studying the processes of Darwinian evolution.

Probably the strongest argument against bona fide machine intelligence is the appeal to Gödel's Theorem, because it seems to provide a logical basis for the supremacy of the human brain over the working of a mechanical device like a digital computer. But, as we have seen, the assumptions underpinning Gödel's result are just the type that cannot be verified in a way that would enable us to conclude that the theorem actually applies in the real-world situation of real minds and brains. Thus we come to the conclusion that if you want to study intelligence in a machine, Gödel's result is no barrier. Moreover, as the arguments at the end of the chapter suggest, it might be a smart move initially to shift your attention from intelligence to life itself, because if you can create life in a computer, then intelligence will surely follow.

Chapter Eight

Time and Time Again

The imposing German philosopher Immanuel Kant held to the view that change is an illusion due to our special human modes of perception. Specifically, in his *Critique of Pure Reason,* Kant states, "Those affections which we represent to ourselves as changes, in beings with other forms of intuition, would give rise to a perception in which the idea of time, and therefore also of change, would not occur at all." For Kant, our faculties of perception do not allow us to perceive things in themselves (*Ding an Sich*). In 1949 Gödel found solutions of Einstein's equations of general relativity that produced theoretical universes in which "no objective lapse of time should be assumed at all." In such a world, time travel into the future or the past is possible in exactly the same way in which we can travel in different directions in space. In Gödel's universe, the distinction between earlier and later is abandoned. The question is whether this mathematically possible universe can be excluded as a candidate for the physical universe on observational grounds. Let's take a harder look at what Gödel achieved and what his cosmological model involves as a first step to addressing this challenging point.

Back to the Future

It is often forgotten that Gödel began his graduate studies in theoretical physics, not mathematics. Inspired by his close friendship with Einstein, in his Princeton years Gödel devoted a considerable amount of time to questions of general relativity theory and the notion of time as an infinite, self-enclosed strip. In 1949 Gödel was asked to contribute to a volume on Einstein's work in the famed series *Library of Living Philosophers*. The resulting paper merged his philosophical interests in the idealism of Kant and his cosmological interests in relativity theory.

Gödel and Einstein at the IAS (ca. 1949)

The foundation upon which the special theory of relativity (STR) rests is that the velocity of light in a vacuum is constant. In particular, this means that every observer will measure the same speed—*regardless* of his or her motion relative to another observer. In particular, this implies that time itself is relative in the sense that the observer who moves relative to a stationary observer ages at a slower rate as seen by that stationary observer. Of course, the moving observer sees nothing unusual about how the clock ticks; it's only the stationary observer watching the moving observer's clock who is surprised to see the moving clock going more slowly than his own stationary clock. This is the phenomenon of *time dilation,* which has now been observed in many laboratory experiments involving particles moving at speeds near the velocity of light.

The general theory of relativity (GTR) extends this idea to argue that an event transpiring in a small region of space and time that has an accelerated frame of reference, not the constant-motion frame of the STR, cannot be distinguished from the same event taking place in a gravitational field. Here *accelerated* means relative to the "resting" position of an observer. If, however, we wish to describe the world in spatial and temporal dimensions, regarding gravity as a curvature, or "warping," of space and time, then the central postulate of GRT is that events must unfold in a four-dimensional space–time continuum (it is usually written *spacetime*)— something very difficult to envision using our limited powers of three-dimensional visualization.

We can approach such a continuum metaphorically, however, by following the lead of philosopher Hans Reichenbach, who regards the temporal dimension as a color. In this scheme, every four-dimensional physical object can be characterized in terms of changes in its color and spatial position. Thus two objects interact only when they have the same color and spatial location. Physical objects of different colors interpenetrate without influencing each other. For example, a swarm of red flies totally enclosed in a red bottle can escape by changing their color to blue.

It's important to note that geometry in the four-dimensional spacetime structure of general relativity corresponds not to everyday euclidean geometry but to what is called *pseudo-Riemannian geometry*. This is a geometry in which neither space nor time is flat; rather, both are curved. This does not mean our three-dimensional space is embedded within another multidimensional space, but only that the laws of euclidean geometry apply in small, local regions of spacetime. The space then undergoes deviations from the flat geometry of Euclid when the regions of space and time become "large." This is analogous to the situation in everyday geometry here on Earth. At any particular location, space seems flat—just as Euclid would have it. But if we take a large enough region, say by flying in a plane at 60,000 feet, then we see the Earth's curvature, which shows that globally Earth's surface is curved, not flat like a table top.

In a noneuclidean geometry, the concept of a straight line as the shortest distance between two points generalizes to the notion of a *geodesic,* as the curve of shortest distance between two points. For instance, the shortest path between two points on a sphere is an arc on a meridian. This is why airplanes take a polar route from New York to Tokyo, so as to conserve fuel and save time. The transition from STR to GRT involves regarding gravity as a curvature of spacetime. Thus the shortest path between two points in spacetime subject to a gravitational field is a curve—but usually one much more complicated than the simple meridian of a sphere.

Here it's useful to recall the words of Leopold Infeld, a long-time colleague and collaborator of Einstein's:

The distinguishing mark of our world's geometry is the gravitational field. Just as a rubber plane can be deformed by the influence of external forces, moving masses deform our spacetime. Geometry and gravitation become synonymous; they are determined by the distribution and speed of different masses.

Hence we can sum up the laws of gravity and inertia in a single sentence: A point mass subject to the laws of gravity alone moves along a geodesic in the spacetime continuum.

Light spreads in the form of photons. In a euclidean view of the situation, the photons in a flash of light spread out cone-like from the flashpoint. But in the spacetime depiction of such an event taking place, say, on the surface of this page of the book, light can spread out only in the plane of the page, because we have removed one coordinate from three-dimensional space to use it to represent time. In this planar example, then, space is depicted as a plane. Consequently, observers looking down on this plane would see light spreading out in a circular fashion from the flashpoint, in much the same way that ripples move on a pond away from the point where a stone is dropped into it.

The result of the expansion of such a circle along its temporal axis (the vertical dimension sitting above the plane) is a cone, the so-called *light cone*. More precisely, it is a double cone. Light rays moving outward from the flash (into the future) constitute the cone's upper half, whereas rays moving inward (from the past) form the lower half of the cone. So the future and the past cones meet in the present moment, the apex of the cone which represents the "now" moment of the light flash on the surface of the page.

Let's now leave aside all perspective drawing and think of the double cone's upper portion alone pro-

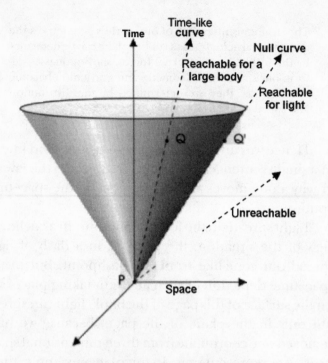

Figure 8.1. The light cone.

jected onto a piece of paper. The cone then appears as a simple triangle. In this setup, three-dimensional space, which was formerly represented by the plane, is now simply a line, a one-dimensional object that has to represent the entire three-dimensional space. Light now spreads out horizontally in this single dimension. But because light travels at a constant speed, it must also be displaced along the vertical time axis. By convention, this displacement is at a 45 degree angle to the vertical line through the cone apex (the origin) and its two axes. The two cone axes thus form a 90 degree angle. The outer surface of such a light cone is called a *null cone*. The entire situation is shown schematically in Figure 8.1.

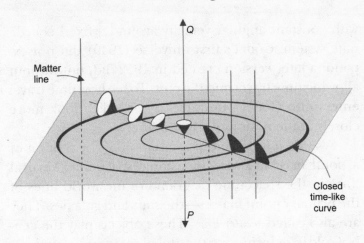

Figure 8.2. Time-like curves.

Gödel's Universe

Null light cones are crucial for the universe Gödel created. Their chief importance lies in how they determine the causal connection between points of spacetime. If a point *P* can be linked to a point *Q* through a null curve directed into the future (a time-like curve), then it is possible for a signal to move from *P* to *Q* (but usually not the opposite). This is the local interpretation of what the null cone means. When we consider it globally, however, in a curved spacetime, the influence of a gravitational field allows temporal world lines to close in on themselves, returning to the same point where they started. Gödel's universe contains just such closed loops in time of the sort depicted in Figure 8.2.

In the 1949 volume celebrating Einstein's seventieth birthday, Gödel presented work that sparked research aimed at finding exact solutions to Einstein's field equations in GTR, solutions that were more complex than any previously known. Gödel's solution regards the entire mass of the universe as an incompressible, perfect fluid. In this model the universe rotates with constant angular velocity around a fixed coordinate system. Gödel's first universe (1949) did not expand; a later version created in 1952 did, but without allowing time travel into the past. To see how time travel enters into Gödel's universes, we need to look more closely at the notion of a world line.

A *world line* is the representation of the motion of a point mass in four-dimensional spacetime. Vertical lines in the spacetime diagram are the world lines of the primary point masses—the stars and galaxies. They are also called *matter lines*. These objects play the central role in determining the global spacetime structure, because they create a strong gravitational field. Similar to the situation with magnets, the geodesic lines for the motion of the particles bend around these point masses into closed time-like curves. The diagram in Figure 8.2 shows the rotational symmetry around the central matter line from point P to point Q.

Matter lines are ordered in time; they have a well-defined notion of succession. A future-oriented curve runs from P to Q, so a time traveler starts from point P and travels along the matter line toward Q. From our usual perspective, the traveller cannot normally go back from Q to P. But in Gödel's universe, closed time-like matter lines are indeed possible. All such world lines correspond to accelerated motion. Thus a trip along such a world line is possible only at great speeds—for instance, with the help of very powerful rockets. Here's how Gödel himself explained things:

> Every world line of matter occurring in the solution is an open line of infinite length, which never approaches any of its preceding points again; but there also exist closed time-like lines. In particular, if P, Q, are any two points on a world line of matter, and P precedes Q on this line, there exists a time-like line connecting P and Q on which Q precedes P; i.e., it is theoretically possible in these

worlds to travel into the past, or otherwise influence the past.

But all this is possible only at great speeds, speeds that would require rockets so powerful that the energy consumption of entire planets would be needed to propel them. In the words of theoretical physicist Heinz Rupertsberger,

> For time travel, speeds exceeding 70 percent of that of light are needed; the required energy is massive. This becomes clear when we imagine the earth as a rocket, its matter serving as propellant that is ejected with the speed of light. Very roughly estimated, for a trip 100 years into a matter world-line's past—a trip in which a hundred years thus go by for the traveler—at least enough earthly material is consumed for the earth to be a sphere with a 6-meter radius at the trip's end.

Such travel into the past naturally calls into question the universe's causal structure. Nevertheless, with the aid of the solution to the equations of GTR created by Dutch cosmologist Jan de Sitter, the famed mathematical physicist Hermann Weyl pointed to the basic principle:

> In a world whose coherence is defined by four-dimensional numerical space, it is easy to construct a metric field in such a way that if the cone of the passive past emerging from a starting point is followed far enough backward, the cone ends up with its own interior overlapping starting point itself. This leads to the most gruesome possibilities of doppelgängers and self-encounters.

Einstein's cylinder and de Sitter's hyperboloid both possess this double border of past and future. However, in the mass-filled world of cylinders, the backward-extended past cone overlaps with itself ad infinitum. It can thus happen that we view different images of the same star in the heavens, representing different epochs in its life separated by vast expanses of time. In such a world, "ghosts" of the long past mingle with us; in de Sitter's hyperbolic world, this self-overlapping of the null cone does not occur.

Presently there is thus no compelling physical reason for excluding acausal behavior in solutions to Einstein's field equation. At most, we might speak of undesirable behavior. As Hermann Weyl indicates, this type of problem led to Stephen Hawking's definition of stable causality in 1963. In Hawking's view, a universe is called *causal* when no closed time-like lines emerge from the introduction of well-defined, extended light cones.

A further set of problems emerging from Gödel's solution to the field equations of GTR led to investigations of singularities in cosmological models. A long space-time curve that closes back on itself at infinity represents such a singularity, which can lead to a black hole or be at the universe's beginning—that is, at the "Big Bang." This was proposed in the "singularity theory" created by Stephen W. Hawking and Roger Penrose in the 1970s.

Gödel's solution thus made causality the center of a new debate in physics. Stepping into the chain of causation is, of course, possible only when the result does not destroy its own cause. In this sense, we might say that time travel takes place in some region beyond causality. It is in this way that we can avoid time travel paradoxes like going back in time to kill your own father before your conception.

Gödel's work on cosmology seems to have very little direct connection to his mathematical work on the limits to deductive reasoning, other than that both appeal to his Platonic view of the way things are—in both the mathematical and the physical realm. But if cosmology seems something of an aberration, our next topic is quintessential Gödel. Just as the work on mind and mechanism revealed the importance of incompleteness as a way of distinguishing the purely logical and deductive from the intuitive and inductive, the subsequent development of the theory of computation has led to new insights into incompleteness that give us a totally novel way of looking at what Gödel's results mean. Moreover, by viewing incompleteness from a computer science point of view, we see new ways to extend Gödel's thinking into such areas as information theory and randomness that Gödel himself never explored. The next chapter takes a quick look at some of these extensions.

Chapter Nine

The Complexity
of Complexity

One way to interpret Gödel's incompleteness result is to say that within any sufficiently strong formal system of deductive logic, there are statements that are too complex to prove using the logical operations of that system. Of course, the notion of "complexity" is an informal one, just as is the notion of "truth." As we saw early in our discussion, Gödel managed to find a way to formalize the informal idea of truth, replacing it with the concept of proof. In this same manner, researchers since Gödel's time (most notably Gregory Chaitin of the IBM Research Laboratories) have managed to find a clever way to formalize the idea of complexity. What is perhaps surprising is that, although on the surface "truth" and "complexity" don't necessarily seem to have much to do with each other, it turns out that they have *everything* to do with each other. In fact, in a certain sense, truth is simply a subset of complexity. Let's see why.

Being Arbitrary

The replacement of the planet Earth by the Sun as the center of heavenly motions is widely (and rightly) seen as one of the great scientific paradigm shifts of all time. But what is often misunderstood is the reason why this Copernican "revolution" eventually carried the day with the scientific community. The commonly held view is that Copernicus's heliocentric model vanquished the competition, especially the geocentric view of Ptolemy, because it yielded better predictions of the positions of the celestial bodies. In actual fact, the predictions of the Copernican model were a little *worse* than those obtained via the complicated series of epicycles and other curves that constituted the Ptolemaic scheme, at least to within the accuracy available using the measuring instruments of the time. No, the real selling point of the Copernican model was that it was much *simpler* than the competition yet still gave a reasonably good account of the observational evidence.

The Copernican revolution is a good case study in how to wield Ockham's Razor to slit the throat of the competition: When in doubt, take the simplest theory that accounts for the facts. The problem is that it's not always easy to agree on what is "simple." The notion of simplicity, like truth, beauty, and effective process, is an intuitive one, calling for a more objective characterization—that is, formalization—before we can ever hope to agree about the relative complexities of different theories.

Basically, simplicity = economy of description. To illustrate this claim, suppose you wanted to put a new tile floor in your bathroom and were considering the two tile patterns shown in Figure 9.1. If you were trying to describe the candidate patterns over the telephone

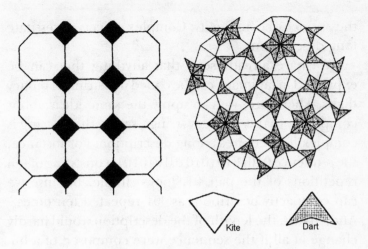

Kite   Dart

Figure 9.1. Two floor tilings.

to a friend, the chances are that you'd have little difficulty conveying a clear picture of the first pattern by
using a very compact description. For example, you
might say, "Alternating columns of large white octagons
and small black diamonds." To write down this description, you would have to use 69 keystrokes on a computer keyboard (including spaces and punctuation).
Moreover, the same description works to describe the
pattern regardless of how large your bathroom happens to be. And with a little thought, it would probably
be possible to create a good description in even fewer
keystrokes. On the other hand, despite the fact that the
second pattern also tiles the floor completely with two
primitive figures (in this case, the kite and the dart), it
seems hard to find a short description that accurately
describes what the overall pattern really looks like. The
problem is that although there seems to be some kind
of structure in this "Penrose tiling," it's hard to capture that structure in a condensed, easy-to-express way.
Thus the Penrose tiling seems intuitively more complex, or complicated, than the first pattern. Here is

the essence of complexity: Complex objects necessitate long descriptions.

We have already seen that anything that can be expressed in language can be coded in a string of binary digits. Therefore, we can apply the same ideas about complexity to these binary strings, saying that a string is complex if it requires a long description. For instance, the sequence 01010101010101010101, consisting of ten repetitions of the pair 01, looks simple, because we can compactly describe it as "01 repeated ten times." And, in fact, the length of the description would hardly change at all if the sequence were composed of a billion or a trillion such pairs. For such a well-patterned sequence, the description is far shorter than the length of the sequence itself. By way of contrast, the string 00110110110010001011, which is exactly the same length as the first one, seems complicated; it exhibits no readily identifiable pattern. Its shortest description appears to be just the sequence itself, written out digit by digit. Thus we intuitively feel that it is complex, just as we regard the first sequence as simple.

In 1964 Ray Solomonoff, a researcher at the Zator Corporation, published a pioneering article in which he presented a scheme to measure objectively the complexity of a scientific theory. He based his idea on the premise that a theory for a particular phenomenon must encapsulate somehow the available observational data characterizing that phenomenon. With this idea in mind, Solomonoff proposed to identify a theory with a Turing machine program that, given a description of the experimental setup as input, would produce the empirical observations as the output. Solomonoff argued that the complexity of a theory could be taken to be the "size" of the shortest such program, measured perhaps by the number of keystrokes needed to type the program or (what effectively amounts to the same thing) the number of bits or bytes needed to express the program in ASCII code.

Using this definition of the complexity of a scientific theory, Solomonoff anticipated an observation made later by the mathematician Gregory Chaitin and by the mathematician and philosopher René Thom, who both noted that the point of a scientific theory is to reduce the arbitrariness in the observational data. A good theory is one that somehow compresses the data, enabling us to describe compactly what's known about the phenomena of concern. On the other hand, if a program (theory) that reproduces the observational data is not appreciably shorter than the actual data itself, then the theory is basically useless. We might just as well account for the observations by writing them down directly, and we don't need a theory to do that. Thus if the *shortest possible* program that reproduces a set of observations is no shorter than a mere listing of the observations themselves, we are justified in calling the observations "random" in the sense that there are no compact laws, or rules, by which the observations can be either predicted or explained—that is, constructively generated by some algorithm or program.

Because this idea lies at the heart of the modern theory of algorithmic complexity, let's try to hammer home the point by using the Chocolate Cake Machine scenario given in Chapter 1. The universe of possible observations is our old Platonic friend consisting of all possible chocolate cakes. Consequently, the experimental circumstances are just the ingredients needed to make any kind of chocolate cake. Now suppose we make an observation in this universe, and our measuring (that is, tasting) apparatus records something that satisfies the description of a *Sachertorte*. The observational data of this experiment, then, are just a description of everything needed to pick out a *Sachertorte* from among the myriad other inhabitants of the chocolate cake world. To write down such a description involves using a certain number of words and punctuation marks, so the description has some length.

Now suppose we want to create a "theory" of the observational evidence—that is, we want a theory of *Sachertorte*. What this means is that we want to write a program—a recipe—that can be processed by the CCM and that will single out a *Sachertorte* from the universe of chocolate cakes. In accordance with the dictates above, the theory will reduce the arbitrariness of what we might have observed in the universe of chocolate cakes by specifying a procedure for making a cake whose description will be in exact agreement with what in fact we have actually observed—the description characterizing *Sachertorte*.

If our recipe is to be a good theory of *Sachertorte*, it must be able to single out a *Sachertorte* from among the universe of all chocolate cakes. And it must do so in a more efficient manner than by merely listing all the features that characterize a *Sachertorte*. Even from the point of view of practical cookery, any recipe that lists only descriptive characteristics of *Sachertorte* (for instance, that it has a layer of chocolate sponge cake, that the cake is covered with glazed chocolate icing, and that the cake and the icing are separated by apricot jam) would be pretty useless. Clearly, we would prefer to have a recipe for actually making the cake. Ideally, in fact, we would like to have the simplest possible such recipe, because that recipe would presumably involve the least work in the kitchen. The length of this simplest possible recipe for making a cake whose description agrees with that of a *Sachertorte* is what we call the *complexity* of *Sachertorte*.

To anyone who's ever tried baking something from the chocolate cake universe, it should be clear from these arguments that some cakes are just more complex than others. And in perfect harmony with this idea of cake complexity, most chefs would probably agree that a chocolate cupcake and a *Sachertorte* are of quite different levels of complexity: The shortest possible recipe for cupcakes is far shorter than the shortest recipe for a *Sachertorte*. And it seems likely that both of these kinds of chocolate cake have complexity less than that of, say, a chocolate hazelnut cake.

But we began this discussion of simplicity and complexity by arguing that simplicity and complexity are both directly related to pattern and structure. Therefore, it's reasonable to say that if something is complex, then it is relatively unstructured or, more prosaically, it is without pattern. Following this line of thought leads immediately to a question: What about randomness? Does randomness correspond in any meaningful sense to a complete lack of structure? Because we already know that everything describable can ultimately be coded by a number, let's leave the world of chocolate cake and go back to look at this question in the universal universe of abstract numbers.

Randomly Speaking

At about the same time Solomonoff was developing his ideas about the complexity of scientific theories, Gregory Chaitin was enrolled in a computer-programming course being given at Columbia University for bright high school students. At each lecture the professor would assign the class an exercise that required writing a program to solve it. The students then competed among themselves to see who could write the shortest program that solved the assigned problem. This spirit of competition undoubtedly added some spice to what were otherwise probably pretty dull programming exercises, but Chaitin reports that no one in the class could even begin to think of how to prove that the weekly winner's program was really the shortest possible.

Even after the course ended, Chaitin continued pondering this shortest-program puzzle, and he eventually saw how to relate it to a different question: How can we measure the complexity of a number? Is there any way that we can objectively claim π is more complex than, say, $\sqrt{2}$ or 759? Chaitin's answer to this question ultimately led him to one of the most surprising and startling mathematical results of recent times.

In 1965 Chaitin, now an undergraduate at the City University of New York, arrived independently at the same bright idea as Solomonoff: Define the complexity of a number to be the length of the shortest program for a universal Turing machine that will cause the machine to print out the number. Using this idea, Chaitin came up with the following complexity-based definition of a random number: A number is *random* if the shortest program for calculating the number is not appreciably shorter in length than the number itself. To express this another way, we can say that a number is random if it is maximally complex. Here, of course, we take the length of a number or program to be the number of binary digits needed to write down that number or program. With this definition, a number such as $\pi = 3.14159265\ldots$ is not random, because arbitrarily many digits of π can be generated using any of a number of known programs of fixed length. Nevertheless, an infinitely long number like π is certainly more complex than a simple number of finite length like 47; we can always use a program like "PRINT 47" to generate the latter quantity. And this shortest program for 47 is quite a bit shorter than the shortest program that will successively crank out the digits of π.

If something as complicated looking as π isn't random, do random numbers really exist? Or does Chaitin's definition define an empty set? The surprising fact is that nearly *all* numbers are random! To see why, think about what it means for a number to be nonrandom. By definition, a number is nonrandom if it can be produced by a computer program whose length is significantly shorter than the length of the number itself. Suppose we consider all numbers having length n, that is, all binary strings of n digits. Each of the n digits can be either 0 or 1, so there are a total of 2^n numbers of length n. Let's compute the fraction of these numbers having complexity less than, say, $n - 5$. That is, we're looking for all numbers of length n that can be produced by a computer program that can be coded in no more than $n - 5$ bits.

Because our interest is in all computer programs that can be coded in no more than $n - 5$ bits, we could actually list each one of these programs. For example, there is one program of length zero (the empty program consisting of no instructions), two programs of length 1 (the single-element strings 0 and 1), four programs of length 2 (the strings 00, 01, 10, and 11), and, in general, 2^k programs of length k. Counting all possibilities, there are a total of $1 + 2 + 4 + \cdots + 2^{n-5} = 2^{n-4} - 1$ such programs having length $n - 5$ or less. Consequently, there are at most this many numbers of length n whose complexity is less than or equal to $n - 5$, because at best each of these programs can produce an output corresponding to an actual number of length n. But we have seen that there are a total of 2^n numbers of length n. Therefore, the proportion of these numbers having complexity no greater than $n - 5$ is at most $(2^{n-4} - 1)/2^n \leq \frac{1}{16}$.

Thus we see that no more than one number in sixteen can be described by a program that's at least 5 bits shorter than the number itself. Similarly, no more than one number in five hundred can be produced by a program 10 or more bits shorter than the number's length—that is, its complexity is 10 or more units away from being random. Using this kind of argument and letting $n \rightarrow \infty$, we can fairly easily prove that the set of real numbers having less than maximal complexity forms an infinitesimally small subset of the set of all numbers. In short, almost every real number is random, because there exists no program that produces the number and is shorter than the trivial program that just prints the number itself. Now let's look at this shortest-program business in a little more depth.

The starting point for Chaitin's remarkable results is the seemingly innocent query "What is the smallest number that cannot be expressed in words?" This statement seems to pick out a definite number. Let's call it U for "unnameable." But thinking about things for a moment, we see that there appears to be something fishy about this labeling. On the one hand, we seem to have just described the number U in words. But U is supposed to be the first number that *cannot* be described in words! This paradox seems to have originally been suggested to Bertrand Russell by a certain Mr. G. G. Berry, a Cambridge University librarian.

Just as Alan Turing had to formalize the intuitive notion of a what is a "computation," the Berry Paradox contains its own unformalizable notion, the concept of denotation between the terms in its statement and numbers. As he arrived at his results on randomness, complexity, and the limitations of rule-based knowledge, part of Chaitin's insight was to see that the way around this obstacle was to shift attention to the phrase "the smallest number not computable by a program of complexity n." This phrase *can* be formalized, specifying a certain computer program for searching out such a number. What Chaitin discovered was that no program of complexity n can ever produce a number having complexity greater than n. Therefore, the program of complexity n can never halt by outputting the number specified by Chaitin's phrase. This fact constitutes an algorithmic complexity version of the unsolvability of the Halting Problem.

More generally, this result shows that even though there clearly exist numbers of all levels of complexity, it's impossible to prove this fact. That is, given any computer program, there always exist numbers having complexity greater than that program can generate. In the words of physicist Joseph Ford of the Georgia Institute of Technology, "A ten-pound theory can no more generate a twenty-pound theorem than a one-hundred-pound pregnant woman can birth a two-hundred-pound child." Speaking somewhat informally, Chaitin's Theorem says that no program can calculate a number more complex than itself. In the cake world, we could loosely interpret Chaitin's Theorem as saying that you can't make a *Sachertorte* from the recipe for chocolate cupcakes. The cupcake recipe is just too simple to generate anything more complicated than cupcakes. Or, equivalently, the operations needed to make a *Sachertorte* are too complex to be carried out with the limited repertoire of steps and actions specified in the recipe for cupcakes.

The implication of Chaitin's Theorem is that for sufficiently large numbers N, it cannot be proved that

a particular string has complexity greater than N. Or
(what is the same thing) there exists a level N such
that no number whose binary string is of length greater
than N can be proved to be random. Nevertheless, we
know that almost every number is random. We just can't
prove that any *given* number is random. Here's a quick
proof of this surprising and important fact.

Let's take an arbitrary, but fixed, number n. By
the arguments above, the likelihood is overwhelmingly
high that this arbitrarily selected number is random.
Suppose we want to prove that n is indeed "typical" in
this sense. Let's assume there did exist a program P
that checks that n can be generated only by a program
longer than P. As long as we select our number n to
be sufficiently large, the existence of such a program
P would constitute a way of proving that n is random.
Let's show why no such program P can possibly exist.

First of all, we use P to generate all programs of
length 1, length 2, and so on. Some of these programs
will actually be proofs that the number n cannot be generated by programs as short as P. But for these proofs
we could have the program P print out the number n,
in effect generating it. Thus P will have generated a
number that it is too short to generate. This contradiction leads us to conclude that no such program P
exists. Consequently, it's impossible to prove that our
arbitrarily selected number n is random—despite the
fact that nearly all numbers really are random.

The essence of the difficulty in proving that any
particular number is random lies in the fact that each
digit in a random number carries positive information,
because it cannot be predicted from its predecessors.
Thus an infinite random sequence contains more information than all our finite human systems of logic put
together. But because nearly all real numbers consist

of an infinite, nonrepeating sequence of digits, we find that nearly all numbers are in fact random. Nevertheless, verifying the randomness of any particular such sequence is beyond the powers of logical proof. Looking at the problem in another way, in order to write down an "arbitrarily long" patternless sequence, we need to give a general rule for each element of the sequence. But then this rule is shorter than suitably large sections of the sequence, so the sequence can't really be random after all! Chaitin's Theorem also gives us another perspective on Gödel:

Gödel's Theorem—Complexity Version

There exist numbers having complexity so great that no computer program can generate them.

Chaitin's Theorem says that if we have some program, there always exists a finite number t such that t is the most complex number our program can generate. Nevertheless, we can clearly see that numbers having complexity greater than t exist. To construct the binary string for one, simply toss a coin a bit more than t times, writing down a 1 when a head turns up and a 0 for tails.

It is thought-provoking to consider the degree to which Chaitin's result imposes limitations on our ability to find or create scientific theories, or laws, compressing our observations of natural and human phenomena. Suppose K represents our best present-day knowledge about mathematics, physics, chemistry, and all the other sciences, whereas M denotes a UTM whose reasoning powers equal those of the smartest and cleverest of human beings. Then we can estimate the number t in Chaitin's Theorem as

$$t = \text{complexity } K + \text{complexity } M + 1 \text{ million}$$

where the last term is thrown in to account for the overhead in the program of the machine M. To estimate the complexities of K and M, the logician and science fiction writer Rudy Rucker has offered the following argument. First of all, suppose that the knowledge in around 1000 books suffices for K. An average-sized book like this one takes around 8 million bits (1 million bytes) to express in ASCII code, so the total complexity of K comes to a number in the neighborhood of 1000×8 million $= 8$ billion. This is as good an estimate as any for the complexity of K. As to M, using a similar argument, Rucker suggests that it should be possible to characterize everything that we need to know about the UTM with the information contained in another 1000 average-sized books. If so, then the complexity of M also amounts to about 8 billion. Thus we conclude that t is certainly less than 16 billion.

The bottom line is that if any worldly phenomenon generates observational data having complexity greater than around 16 billion, no such machine M (that is, human) will be able to prove that there is some short program (that is, theory) explaining that phenomenon. Thus, recalling René Thom's idea of scientific theories as arbitrariness-reducing tools, Chaitin's work says that our scientific theories are basically powerless to say anything about phenomena whose complexity is much greater than 16 billion. But note that Chaitin's Theorem also says that the machine will never tell us that there does *not* exist a simple explanation for these phenomena, either. Rather, it says that if this "simple" explanation exists, we will never understand it—it's too complex for us! Complexity 3 billion represents the outer limits to the powers of human reasoning; beyond that we enter the "twilight zone," where reason and systematic analysis give way to intuition, insight, feelings, hunches, and just plain dumb luck.

Although we didn't emphasize the point in discussing Turing machines, one of the things that we could use such a computing device for, at least in principle, would be to search for all the true statements of arithmetic. From Turing's solution of the Halting Problem and its equivalence to Hilbert's Decision Problem, as well as from the faithful correspondence between Turing machines and formal systems, we know that there is no program that will ever print out all the true statements of arithmetic. Just for fun, let's rephrase Gödel's Theorem in these terms. Call a program P *correct* if it never lists a false statement of arithmetic. Then a truth omitted by P is a true statement of arithmetic not listed by P. With these definitions, we have

---

Gödel's Theorem—Computer Program Version

There is a computer program $\tilde{P}$ such that if P is a correct program, then $\tilde{P}$ applied to P yields a truth omitted by P.

Gödel's result can also be obtained from what amounts to a completely straightforward, almost trivial corollary of Chaitin's Theorem on complexity, again appealing to the matchup between Turing machines and formal systems. Suppose we are given a consistent formal system F. Then Chaitin's Theorem says that there are numbers having complexity greater than F can prove. In other words, F is incomplete. This amounts to a one-line proof of Gödel's Theorem—Complexity Version, which was stated earlier in terms of Turing machine programs.

Remarkable as it is, Chaitin's Theorem on complexity is only the appetizer to a main course whipped up recently by Chef Chaitin, showing that there are arithmetical facts that completely elude the bounds of finitary rules of reasoning. What Chaitin proved is that although we can clearly state these simple propositions, *for all intents and purposes their truth or falsity might as well be settled by flipping a coin; they are completely and forever beyond the bounds of the human mind ever to resolve definitively.* Let's complete our tour of the ins and outs of Gödel's Theorem with an account of this stunning result.

The Tenth Problem

If asked to name the top ten theorems of all time, just about every mathematician would reserve a place somewhere on the list for the Pythagorean Theorem, which relates the lengths of the sides of a right triangle like the one shown in Figure 9.2. The Pythagorean Theorem says that if a and b are the lengths of the short sides of such a triangle, and c is the length of the hypotenuse, then the equation $a^2 + b^2 = c^2$ linking the quantities a, b, and c always holds.

The Pythagorean Theorem is an example of a polynomial equation in three variables: a, b, and c. The solution to this equation is a set of values for a, b, and c that satisfy the equation, such as like $a = 1$, $b = 2$, $c = \sqrt{5}$. Of special mathematical interest are the so-called *Diophantine equations,* polynomial equations for which we demand that the solutions be in whole numbers (unlike the example just given, which involves the noninteger quantity $\sqrt{5}$). To illustrate, the three quantities $a = 3$, $b = 4$, $c = 5$ form an integer solution to the equation. Thus when we consider the equation linking a, b, and c in the Pythagorean Theorem and admit only integer solutions like this, then we are thinking of it as a Diophantine equation. Consequently, the term

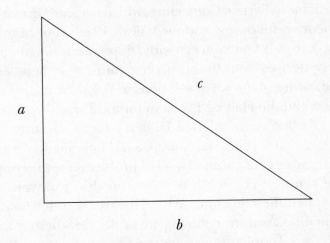

Figure 9.2. The Pythagorean Theorem: $a^2 + b^2 = c^2$.

Diophantine refers more to the character of the set of solutions we're interested in finding than it does to the equation itself.

The number of solutions of a given polynomial equation may vary from finite to infinite, depending on whether we regard it as a Diophantine equation. For example, the Pythagorean Theorem's equation can be shown to have an infinite number of both noninteger and integer solutions. On the other hand, the equation $a^2 + b^2 = 4$ has only the four integer solutions $a = \pm 2$, $b = 0$ and $a = 0$, $b = \pm 2$, but it has an infinite number of noninteger solutions (for example, a any real number between $+2$ and -2, $b = \sqrt{4 - a^2}$). Regarded as a Diophantine equation, then, this equation has a finite solution set. But thought of as a general polynomial equation, it has an infinite number of solutions. (Incidentally, the matter of whether the Diophantine equation $x^n + y^n = z^n$ has *any* positive solution when n is an integer greater than 2 was one of the most celebrated unsolved problems in mathematics, the famous Fermat Conjecture, which became Fermat's Theorem following Andrew Wiles's 1993 proof of this conjecture.) Our concern with Diophantine equations here derives from the surprising connection between the nature of the set of solutions to Diophantine equations and the Halting Problem for a UTM.

Earlier we mentioned Hilbert's famous lecture to the 1900 International Congress of Mathematicians in Paris, where he outlined a set of problems for the coming century. The tenth problem on this list involved Diophantine equations. What Hilbert asked for was a general algorithm enabling us to decide whether an arbitrary Diophantine equation has any solution. For future reference, note carefully that Hilbert did not ask for a procedure to decide whether the solution set was infinite, but only for an algorithm to determine whether there is *any* solution.

It turns out that there exists an algorithm for listing the set of solutions to any Diophantine equation. In principle, all we have to do to decide whether the solution set is empty is to run this program, stopping the listing procedure if no solution turns up. The difficulty is that it might take a very long time (like forever!) to decide whether a solution will appear. For instance, the first integer solution of the simple-looking Diophantine equation $x^2 - 991y^2 - 1 = 0$ is

$$x = 379516400906811930638014896080$$
$$y = 12055735790331359447442538767$$

How long would you be willing to punch keys on a calculator waiting for that pair to pop up? This example shows that to solve Hilbert's Tenth Problem, we can't rely on a brute-force search for the first solution; a solution might not exist, or it might be so large that we'd get tired of looking. In either case, a direct search gives no guarantee of ever coming up with the correct answer about a particular equation's solvability. We need to do something a bit more clever.

While he was an undergraduate student in mathematics at the City College of New York shortly after World War II, Martin Davis read in one of his teacher Emil Post's articles that Hilbert's Tenth Problem "begs for an unsolvability proof." Following many years of effort by Davis, along with major contributions by Julia Robinson and Hilary Putnam, the whole issue boiled down to showing that if there existed even one Diophantine equation whose solutions behaved in a particularly explosive fashion, then Hilbert's Tenth Problem would be resolved in the negative. Unfortunately, no one was able to prove the existence of such an object until 1970, when Yuri Matyasevich, a 22-year-old mathematician at the Steklov Mathematical Institute in Leningrad, found an example of the elusive type of equation. Amusingly, Matyasevich made crucial use of the famous Fibonacci sequence of numbers in constructing his solution to Hilbert's problem. This sequence, which was originally introduced by Leonardo of Pisa in 1202 to explain the explosive growth of a rabbit population in the wild, showed that apparently the well-known procreation habits of rabbits give rise to just the kind of rapid growth Matyasevich needed to create his equation and thus definitively resolve in the negative yet another of Hilbert's conjectures.

An interesting corollary of Matyasevich's proof is that there exists a polynomial such that as the variables a, b, c, and so on take on all positive nonnegative integer values, the positive values of the polynomial itself are exactly the set of prime numbers. To illustrate concretely this curious result, researchers have produced such a polynomial involving 26 variables, whose positive values coincide with the set of prime numbers.

By now the reader should be highly sensitized to the connection between negative solutions to decision problems and Gödel's Theorem. Thus before continuing our pursuit of the connection between Diophantine equations and the Halting Problem, let's pause to give Gödel his due—again.

Gödel's Theorem—Diophantine Equation Version

There exists a Diophantine equation having no solution— but no theory of mathematics can prove this.

Part of Matyasevich's celebrated solution to Hilbert's Tenth Problem focused attention on the little-known fact that any computation can be encoded as a polynomial. In other words, for every Turing machine there exists an equivalent Diophantine equation, and the properties of the solutions of this equation mirror precisely the computational capacities of the corresponding Turing machine. Thus, not only are formal systems and Turing machines abstractly identical, so are Turing machines and Diophantine equations. Chaitin's recent results on the randomness of arithmetic make heavy use of this fact within the context of a generalized type of Diophantine equation called an *exponential Diophantine equation*.

The standard kind of Diophantine equation that we saw with the Pythagorean Theorem involves variables such as a, b, and c that can be raised to some integer power (2 in the Pythagorean case). In an exponential Diophantine equation, the variables can be raised to the power of other variables. To illustrate the idea, $a^b + 5c^3 - d^{3e} = 0$ is such an equation, where we see the variables a and d raised to the power of other variables. Chaitin's work uses a family of such equations, each member of the family being singled out by the value of a single variable k (called a *parameter*) that ranges through the natural numbers. The equation $ka^2 + 3b^c = 0$ is an example of this kind of infinite family, because as we let the parameter k take on the values $k = 1, 2, 3, \ldots$, we obtain the individual exponential Diophantine equations $a^2 + 3b^c = 0$, $2a^2 + 3b^c = 0$, $3a^2 + 3b^c = 0$, and so on.

Chaitin's work is based on a result of James Jones and Matyasevich's to the effect that it is possible to find such a family of exponential Diophantine equations with a single parameter k, such that the equation has a solution for a given value of k if and only if the kth computer program for a UTM (the program whose Gödel number is k) ever halts. Of course, this result shows clearly the complete equivalence between Hilbert's Tenth Problem on the solvability of Diophantine equations and the Halting Problem. Now let's see how Chaitin extended these ideas to show that there is just nothing certain in life—even in the world of numbers.

Omega Is the End

Buried deep within the Theoretical Physics Division of the IBM Research Laboratories in Yorktown Heights, New York, is a broom-closet–sized office, whose spartan furnishings consist of a bare desk, three empty bookshelves, a spotless blackboard, a Monet landscape reproduction on the wall, and a computer terminal. After having spent over 20 years as an IBM salesman, systems engineer, and programmer, Gregory Chaitin now calls this office home. And in 1987 it was from these stark surroundings that Chaitin hurled forth a lightning bolt so electrifying that the editors of the *Los Angeles Times* wrote in their June 18, 1988, editorial that "Chaitin's article makes the world shake just a little." What kind of mathematical result could possibly send a national newspaper into such a state of rapture? Nothing less, it turns out, than a proof that the very structure of arithmetic itself is random. It must surely stand as being about as close to the final word on mathematical truth, proof, and certainty as we'll ever get. Let's see how Chaitin managed to extend Gödel's results to come up with such an astounding conclusion.

Suppose we have a UTM and consider the set of all possible programs that can be run on this machine. As we already know, every such program can be labeled by a string of 0s and 1s, so it's possible to "name" each program by its own personal ID number. Consequently, it makes sense to consider listing the programs, one after the other, and to talk about the kth program on the list, where k ranges through the positive integers. Now consider the following question: "If we pick a program from the list at random, what is the likelihood that it will halt when run on the UTM?" Or, equivalently, we could start with a fixed program for the UTM and ask the same question for an input string that's random. It turns out that this question is intimately tied up with the solvability of Diophantine equations, leading eventually to Chaitin's remarkable result.

The key step in Chaitin's route to ultimate randomness was to consider not whether a Diophantine equation has *some* solution but the sharper question of whether the equation has an infinite or a finite number of solutions. The reason for asking this more detailed question is that the answers to the original query are not logically independent for different values of k. In other words, if we know whether some solution exists for a particular value of k, then this information can be used to infer the answer for other values of k. But if we ask whether there are an infinite number of solutions, the answers are logically independent for each value of k; knowledge of the finiteness (or not) of the solution set for one value of k gives no information at all about the answer to the same question for another value.

Following this reformulation of the basic question, Chaitin's next step was a real *tour de force*. He proceeded to construct explicitly a particular exponential Diophantine equation family specified by a single parameter k, together with over 17,000 additional variables. Let's call this equation $\chi(k, y_1, y_2, \ldots, y_{17,000^+}) = 0$, using the Greek symbol χ (chi) in Chaitin's honor. From this equation we can form a very special string of binary digits in the following manner: As k successively assumes the values $k = 1, 2, 3, \ldots$, we set the kth entry in our string to 1 if Chaitin's equation $\chi = 0$ has an infinite number of solutions for that value of k, whereas we set the kth entry to 0 if the equation has a finite number of solutions (including no solution). As we already know, the binary string we form via this procedure represents a single real number. Chaitin labeled this number Omega, for the last letter in the Greek alphabet, Ω. And for good reason: The properties of Ω show that it's about as good an approximation to "The End" as the human mind will ever make.

First of all, Chaitin showed that the quantity Ω is an uncomputable number. Furthermore, he proved that any program of finite complexity N can yield at most N of the binary digits of Ω. Consequently Ω is random, because there is no program shorter than Ω itself for producing all of its digits. Moreover, the digits of Ω are both statistically and logically independent. Finally, if we put a decimal point in front of Ω, it represents some decimal number between 0 and 1. When viewed this way, Ω can be interpreted as the probability that the UTM will halt if we present it with a randomly selected program—or, as the probability that a fixed program will halt if presented with a random input. Indeed, Chaitin constructed his equation precisely so that Ω would turn out to be this halting probability.

Thus, whereas Turing considered the question of whether a given program would halt with a given input, Chaitin's extension yields the probability that a randomly chosen program will stop. As an aside, it's worth noting that the two extremes $\Omega = 0$ and $\Omega = 1$ cannot occur. The first case would mean that no program ever halts, the second that every program will halt. The trivial, but admissible, program STOP deals with the first case, and we leave it to the reader to construct an equally primitive program to deal with the second.

But the real bombshell, the one that shook up the editorial staff at the *Los Angeles Times*, is that the structure and properties of Ω show that arithmetic is fundamentally random. To see why, take some finite but "sufficiently large" integer. For example, some number greater than the Busy Beaver function value $BB(12)$ considered in Chapter 1. For values of k larger than this, there is no way to determine whether the kth digit of Ω is 0 or 1. And there are an infinite number of such undecidable digits, each corresponding to the following simple, definite arithmetical fact: For that value of k in Chaitin's equation $\chi = 0$, the equation has either a finite or an infinite number of solutions. But as far as human reasoning goes, which of the two possibilities is actually the case may as well be decided by flipping a coin; it is completely and forever undecidable and hence effectively random.

Thus Chaitin's work shows that there are an infinite number of arithmetical questions with definite answers that cannot be found using any axiomatic procedures; they do not and cannot correspond to theorems in any formal system. The answers to these questions are uncomputable and are not reducible to other mathematical facts. Extending Einstein's famous aphorism about God, dice, and the universe, Chaitin describes the situation by saying, "God not only plays dice in quantum mechanics, but even with the whole numbers." It's fitting to conclude this section with our final tribute to Gödel:

Gödel's Theorem—Dice-Throwing Version
There exists an uncomputable number Ω whose digits correspond to an infinite number of effectively random arithmetical facts.

This work reinforces the message of algorithmic information theory that randomness is as fundamental and as pervasive in pure mathematics as it is in theoretical physics. It also lends further support to "experimental mathematics" and to the "quasi-empirical" view of mathematics, which holds that although mathematics and physics are different, that difference is largely a matter of degree. Physicists are used to working with assumptions that explain a lot of data but might eventually be contradicted by experimental results. Mathematicians don't like this kind of "tentativeness"; they want logical certainty. When a conjecture becomes a theorem, it should stay a theorem—forever! Even after Gödel and Turing showed that Hilbert's dream didn't work, in practice most mathematicians carried on in Hilbert's spirit—that is, more or less as before. Finally, though, the computer is changing the way we do things.

It is easy to run a mathematical experiment on a computer and obtain some result. But it's not always easy to create a proof to explain the result. In order to cope, mathematicians are sometimes forced to proceed in a more pragmatic manner, like physicists. The results on Omega provide a theoretical underpinning for this revolutionary approach to doing mathematics.

Chapter Ten

Window on the Soul

By 1931 Gödel had mastered the art of using precise analysis to forge new paths through the labyrinth of self-referential thinking. His opening up of fresh conceptual horizons was apparent in each stage of his career: the early years, marked by triumphs in mathematics and logic; the second phase, in which he turned to questions of physics in the hope of repeating his earlier successes; and his later years, which were devoted chiefly to the elaboration of philosophical problems.

Gödel's mathematical philosophy was resolutely Platonistic. He assumed that mathematical objects existed in some realm beyond space and time—but that they were no less real because of this. In his words, "We have a certain perception of the objects of quantitative theory, and we also form our ideas of these objects on the basis of something that is directly given." This is a Platonic view of mathematical objects, no doubt about it. For the Platonist, objects are thus intuitively present. By way of contrast, an Intuitionist or Constructivist considers them inventions of the human mind.

The mathematical "realist" like Gödel thus grasps the independently existing mathematical objects with his intuition, and then proves the properties of the objects using logical analysis. Hence mathematical intuition is a means to a cognitive end, not simply a source of mental fictions. As the French mathematician and philosopher René Thom put it, "The voice of reality is in the sense of the symbol." For Gödel, as well, we find a typically Platonist intertwining of an objective concept of reality with a kind of extrasensory perception of abstract, Platonic ideas. For Gödel, there were no more grounds for doubting the existence of the objects of his study than there are for physicists to doubt the reality of the material objects on which they focus their investigation. Such mathematical objects must then exist outside of space and time, and it is thus not surprising that Gödel later came to be interested in ESP, transmigration, and occultism in all its variants.

In Gödel's work, then, we do not invent—we discover. Either we "see" and grasp the mathematical object with our intuition or we do not. But even when we do perceive the objects, the linguistic means for describing them are distinctly limited. Gödel's Incompleteness Theorem can thus be viewed as a kind of "logical pessimism," though one with wide ramifications. For if formal means are too weak to prove all the true propositions that can be stated within even the highly restricted confines of a formal system, then our mental tools are clearly too weak to understand—at least by any formal, deductive means—the highly complex system that is the world at large. For Gödel, this does not mean, however, that we cannot come ever closer to the truth, step by step.

Interestingly, Gödel also had a Intuitionist streak in his mostly Platonic view of mathematics. Formalists and Platonists are diametrically opposed in their view of the question of the reality of mathematical objects.

But their research methods and principles of mathematical argumentation are similar. Although Gödel's mathematical philosophy was Intuitionist, his logical methods were Formalistic and his intellectual instruments Logicist. He was thus not an intuitionist Platonist in the limited sense of Brouwer, who argued for the existence of mathematical objects only if they could actually be constructed. Gödel maintained the much stronger position that objects such as the number π and a level of infinity between the integers and the reals exist independently of whether they can be constructively demonstrated. This is classical Platonism in its purest form.

Gödel's famous theorem is an appeal to the inexhaustibility not only of mathematics but of human intelligence in general. For this reason, the theorem also has a powerful kind of ambivalence. On the one hand, it is our century's most important limitation result, dashing the human dream of complete, contradiction-free knowledge—a dream that had persisted for over two millennia. In setting limits on the human fantasy of omnipotence, Gödel stands in the tradition of Copernicus, Darwin, and Freud. On the other hand, through his discovery of the relative nature of human knowledge, Gödel confirms the triumph and necessity of the human spirit and of human intuition.

Let us conclude our account of Gödel's magnificent achievement by quoting Gödel's view of the unlimited nature of the human mind:

> The human spirit is incapable of formulating (or mechanicizing) all its mathematical intuitions. That is, it is when it has succeeded in formulating a portion of them, precisely this fact needs a new intuitive knowledge, for example the consistency of this formalism.

Enough said!

References

1. Dawson, J. *Logical Dilemmas*. Wellesley, MA: A. K. Peters, 1997. The best general biography on Gödel, his life and work. Dawson was responsible for sorting out and organizing Gödel's papers, and thus had unrivalled access to much of Gödel's unpublished papers and notes. Much of it appears in one form or another in this fine volume.

2. Wang, H. *Reflections on Kurt Gödel*. Cambridge, MA: MIT Press, 1987. Hao Wang was an eminent logician who spoke and corresponded with Gödel over the course of many years. This volume, and the one cited next, constitute a record of those conversations, together with Wang's interpretation and historical summary of Gödel's life and work.

3. Wang, H. *A Logical Journey*. Cambridge, MA: MIT Press, 1996. Continuation of the story told in Reference #2.

4. Kreisel, G. *Kurt Gödel, 1906–1978*. London: Biographical Memoirs of Fellows of the Royal Society, Volume 26, 1980, pp. 148–224. Extensive account of

Gödel's life and work by the well-known Austrian-American logician Georg Kreisel. Much food for thought, especially on the philosophical implications of Gödel's work, along with a lot of interesting historical context.

5. *Collected Works, Vols. 1–3.* S. Feferman et al, eds. New York: Oxford University Press, 1986, 1990, 1995. The definitive source for Gödel's own words—published and otherwise—on logic, philosophy, physics and metaphysics.

6. Hofstadter, D. *Gödel, Escher, Bach: An Eternal Golden Braid.* New York: Basic Books, 1979. Pulitzer-prize-winning volume that put Gödel's name before the general public for the first time. Lots of clever stories illustrating Gödelian incompleteness, together with its implications and connections with art, music and much more.

7. Dawson, J. "Kurt Gödel in Sharper Focus." *Mathematical Intelligencer,* Volume 6, Nunber 4, 1984, pp. 9–17. Popular summary of Reference #1. Easily readable by the layman.

8. Taussky-Todd, O. "Remembrances of Kurt Gödel." *Gödel Remembered,* Salzburg, Austria, July 1983. Naples, Italy: Bibliopolis, pp. 29–41. Olga Taussky-Todd was an eminent number theorist who studied at the University of Vienna at the same time as Gödel. This memoir gives her account of those days, and contains a wealth of interesting information about the intellectual milieu that shaped Gödel's view of mathematics and philosophy.

9. Rucker, R. *Mind Tools.* Boston: Houghton-Mifflin, 1987. Very readable layman's account of a lot of fascinating topics relating to Gödel's work, com-

plexity, Chaitin's Theorem, the infinite and much, much more. Highly recommended.

10. Chaitin, G. *The Limits of Mathematics*. Singapore: Springer, 1998. First-person account of the development of algorithmic information theory. Contains computer programs for actually calculating lower bounds on the information content of bit strings.

Index